Social Data Science Xennials

Gian Marco Campagnolo

Social Data Science Xennials

Between Analogue and Digital Social Research

Gian Marco Campagnolo
Science, Technology and Innovation Studies
University of Edinburgh
Edinburgh, UK

ISBN 978-3-030-60357-1   ISBN 978-3-030-60358-8   (eBook)
https://doi.org/10.1007/978-3-030-60358-8

This Palgrave Macmillan imprint is published by the registered company Springer Nature
Switzerland AG.
The registered company address is: Gewerbestrasse 11, 6330 Cham, Switzerland

Acknowledgements

The key arguments in this book are that social data science is a collective effort and that diversity pays dividends. I indeed owe a lot to people I worked with from within the discipline and beyond. I am grateful to Giolo Fele, who took interest into why, as an "Information Systems" person, I was interested in ethnography and tried to explain me what ethnomethodology is. My research is always deeply indebted to Claudio Ciborra, a leading advocate of the separation of computer science from the mathematical and formal sciences. I also want to thank Robin Williams, for always challenging me to question my existing modes of research.

A big thank goes to statistician and applied mathematician colleagues at the Alan Turing Institute for giving me the opportunity to ask questions on what happens inside the machine. The book is thought as a conversation with key scholars in the social data science debate: Rob Procter, Noortje Marres, Celia Lury, Tommaso Venturini, Michael Castelle, Giuseppe Veltri, David De Roure, Hjalmar Bang Carlsen, Anders Blok. This book is the continuation of our conversations.

I also want to thank the Edinburgh colleagues I had the privilege to work with to conduct the studies presented in this book: Neil Pollock, Hung The Nguyen, Beatrice Alex, Gil Viry and Duncan Chapple.

The early idea of this book emerged as a series of seminars I gave at the Marco Biagi Foundation in Modena in April 2019. Thanks to Tommaso Fabbri and Ylenia Curzi for inviting me. The book began to take shape in Copenhagen in November and December 2019 during my eventful fellowship at the Centre for Social Data Science (SODAS). Thanks to SODAS colleagues for their support.

CONTENTS

1 Social Data Science Xennials   1

2 Phenomenological Extensions   19

3 The Analogue Mapping   35

4 Participative Epistemology   51

5 Ethnography as Data Science   73

References   91

Index   103

LIST OF FIGURES

Fig. 3.1   This is a situational map I developed as part of a study on the
            implementation of enterprise software in the public sector. It is
            meant to show how the situation under scrutiny is populated
            by a variety of other organisations that, to different extents,
            contribute to shaping ERP implementation in a particular
            setting/organisation. There are other local organisations from
            different public service domains and other local government
            organisations from different regions that are engaged in similar
            technological projects and share participation in the social
            world/arena   39

Fig. 4.1   This visualisation represents careers started before 1990.
            Horizontal axis represents time from beginning of career to
            time of the analysis. The vertical axis represents number of
            sequences. Careers are sorted from longer (no 1 at the bottom
            of the data region) to shorter (no 84 at the top).
            (Source: Author)   59

Fig. 4.2   This visualisation represents careers started after 2000.
            Horizontal axis represents time from beginning of career to
            time of the analysis. The vertical axis represents number of
            sequences. Careers are sorted from longer (no 1 at the bottom
            of the data region) to shorter (no 67 at the top) to facilitate
            pattern identification. (Source: Author)   60

CHAPTER 1

Social Data Science Xennials

Abstract This chapter explores the tensions between analogue and digital methods in a processual way, placing social data science within the genealogy of the long-term disciplinary relations between phenomenological sociology, expertise in computer science associated with digitalisation and the narrative positivism linked with the use of statistics in social research. Focusing on what endures as well as on what changes, it discusses the theoretical, epistemological and ontological sensibilities that are involved in a commitment to digital data analysis. Referring to the ESRC Digital Social Research programme and to more recent work by the Alan Turing Institute Interest Group in Social Data Science, it acknowledges a UK-centric take on Social Data Science.

Keywords Social Data Science • Digital sociology • Digital social research • Computational social science

Xennials are the demographic cohorts born in the late 1970s to early 1980s that are described as having had an analogue childhood and a digital adulthood. In this book, I want to use this neologism as a metaphor to describe the intellectual biography of social research scholars who were academically "born" before the data deluge (Anderson 2008), when social science was still analogue, and then adapted to the digital "revolution" in research methods.

G. M. Campagnolo, *Social Data Science Xennials*, https://doi.org/10.1007/978-3-030-60358-8_1

1

As many other Xennials, I started my academic education when Apple entered the consumer portable market and completed it just before big data emerged as a buzzword. As such, I take pride in having been able to undertake social research before and after internet-generated data became a regular appearance in the research design of social research projects and PhD dissertations.

By using the term Xennials, I do not mean literally that the experience of moving between analogue and digital social research should be seen a prerogative of the micro-generation born exactly between 1977 and 1983. There are illustrious examples of more senior scholars who also personify research agendas that are grounded in long-standing sociological interests and have subsequently turned to consider the adoption of computational methods in addition to established social research methods.[1]

There have been many ways to explore the tension between analogue and digital methods (see Beaulieu 2016 for a review) but none of them explore this in a processual way. The purpose of this book is to explore the tension between analogue and digital as part of an evolving research programme and to explore the sequencing of methods within it. The book also responds to a growing demand to place digital research more organically in the context of existing ways of doing social science (Savage 2015: 297). Quoting C. Wright Mills—an author highly regarded by digital research scholars (see Edwards et al. 2013; Savage 2015)—with this book I also want to call for sociologists to address connections between personal

---

[1] I refer for example to work by David Stark and colleagues (Vedres and Stark 2010) and to recent developments in research on organisational routines (Pentland et al. 2012). After having carried out decades of ethnographic field research to study the organisational basis of innovation, Stark and his team have recently started to bring the analytical tools of network modelling onto the terrain of cultural sociology. This move granted the authors opportunities to reflect on the intersection of observation theory and network analysis, suggesting that social scientists should adopt a binocular view and reflect on ways to see the same objectivity that are out with the researcher's own theoretical tradition.

Another example is Brian Pentland's research evolution towards computational methods. Having studied organisational routines through intensive fieldwork that involved collecting interview data, deeply rooted in qualitative research tradition of socio-materiality (Pentland et al. 2011), Pentland turned to adopt sequence and process analysis with an interest for identifying regularities across contexts. His intent to break disciplinary boundaries derives from the observation that 'research grounded in a social science tradition tends to focus on people, while research grounded in an engineering tradition tends to focus on artifacts. However, as people and artifacts become increasingly intertwined in digitized processes and practices, these traditional disciplinary divisions sometimes seem a little outdated' (Pentland 2013).

troubles with digital data analysis and more public issues regarding the role of social science in the digital age (Mills 1959).

Most of the authors who have recently written about digital social research fail indeed to remember—or intentionally omit from their accounts—what they were doing in their research before the digital as if it was an incoherence.[2] This intellectual oblivion contributes to construe the digital as the bearer of epochal change in social research, an epistemological posture that in itself contributes to make the very notion of digital research unsettling for most sociology scholars, historically not prone to a rapidly moving research front (Collins 1994). As we shall see, social data science should be better placed within the genealogy of the long-term disciplinary relations between phenomenological sociology, expertise in computer science associated to digitalisation and the narrative positivism (Abbott 1992) linked with the use of statistics in social research.

THE DISUNITY OF SOCIAL DATA SCIENCE

Technical boosterism is not the only reason for an uptake of digital data analysis in social research that remains on a smaller scale than some early visions initially anticipated (Halfpenny and Procter 2015: 5; Lazer and Radford 2017: 20). The other main reason is the continued ambition in social data science scholarship to speak to multiple audiences at once as in an attempt to define specific parameters around the (inter-)discipline. This is apparent either in monographies and edited collections aimed to address all issues and topics that could be incorporated under a sociology of digital technology (Lupton 2014; Orton-Johnson and Prior 2013) or in works explicitly written for both the social science and the data science audiences (Salganik 2018; Veltri 2019). Some work of the former category is theoretically sound. But it remains conceptual. The conclusions are convincing but rarely put into practice. On the other hand, there are practical guides, where authors provide a wide variety of example-driven accounts that target the skeptics and the enthusiasts, the social scientists as well as the data scientists.

My argument is that both approaches reproduce a rather benign view of a unified social data science. Paradigms are so many and their combination so unique across big data sciences (Bartlett et al. 2018) as well as within social data science that any attempt to construct the (inter-)

---

[2] A similar point is made by Atkinson and Housley's (2003) book *Interactionism: An Essay in Social Amnesia*.

discipline as a straightforward project is bound to remain conceptual and solidify disciplinary borderlands.

This volume follows up recent conceptual scholarship in digital sociology (for an excellent example see Marres 2017) with a more 'confidential', first-person narrative about research methods in this particular form of contemporary interdisciplinarity (Sandvig and Hargittai 2015: 2). It also goes to suggest that a degree of digressions and transgressions (Rheinberger 2011) under the umbrella of social data science is a productive feature.

The book provides a first-hand account from the perspective of an ethnomethodology-trained qualitative social scientist that over time got to approach social problems by using alternative methods also including digital methods. As such, the book is written to intervene in internal discussions within the social sciences about possibilities for collaboration with data science. It is addressed specifically to scholars who, like the author, routinely engage in critical sociological analysis of the digital workplace and find it easier to treat the digital as an object of study. It describes how it happened that the transformation of the workplace taking place in the 10-year arc of a career spent doing fieldwork in the IT sector led to progressively question existing methodological presuppositions in social research and "embrace" new forms of data and methods.

If the goal is to act upon a situation in social data science that Bartlett et al. (2018) describe as '98% computer scientists and 2% sociologists', the editorial politic of producing introductory books on data analytics for social scientists might not be the only possible solution. Another, arguably more promising approach is to consider the theoretical, epistemological and ontological sensibilities that might be involved in a commitment to digital data analysis. Even more important for an audience of empirical social researchers influenced by symbolic interactionism and ethnomethodology is to provide a narrative that is concrete and accountable. This work is a succinct effort in this direction. It targets a specific audience, however large. It does so from a partial, but hopefully well-situated perspective: the first-person perspective of how a particular research agenda has been conducted in the field.

A Processual Perspective on Social Data Science

Differently from most of the rapidly burgeoning literature on digital social research, the bulk of the book is about how a social scientist can gradually get to appreciate digital research methods. The most part of what you will read describes the progression of issues, methods, questions and approaches through which the always coming crisis of empirical sociology (Gouldner 1970; Savage and Burrows 2007) can be approached without necessarily jumping straight onto the digital bandwagon. This aligns with the idea that the focus on social science ideas in the context of the changing character of social phenomena should be as much on what endures as it is on what changes (Housley and Smith 2017).

Continuing with an effort to clearly situate the perspective of this book, the work presented here acknowledges a UK-centric (or rather Edinburgh-centric) take on social data science as well as a particular positioning regarding the above-mentioned genealogies in computer science and the academic social sciences.

A considerable amount of what is defined by the ESRC as Digital Social Research[3] has been inspired by topological network approaches.[4] This is now a well historicised trend, initiated by projects such as Controversy Mapping[5] at Sciences Po Médialab (Venturini et al. 2017; Bouillier 2018), brought forward by the IssueCrawler software project of the Digital Methods initiative led by Richard Rogers at the University of Amsterdam (Rogers and Marres 2000; Marres and Moats 2015) and continued in the UK by projects such as 'Reading the Riots' (Procter et al. 2013) and the Collaborative Online Social Media Observatory (Burnap et al. 2014). The digital social research I will be talking about in this book is not part of this lineage.

This book comprises work that goes to show that there is a wide variety of existing social research approaches—not least phenomenological concepts themselves—that can help account for the relational and spatially distributed nature of social action. These approaches—discussed in detail in Chap. 2—show a deeper understanding of the subjective, institutional, social and political processes whereby research can be conducted on a larger scale without the need of digitally produced network graphs.

Also due to the long-standing linkages between Edinburgh and the Chicago School, crucially represented by the shared intellectual legacy of Erving Goffman, the work I will be discussing here embodies an interest for process, that is, the unfolding of phenomena over time (Abbott 1992) and it involves the application of innovative social science methods such as

[3] https://esrc.ukri.org/files/research/research-and-impact-evaluation/dsr-report-executive-summary/.

[4] Moats and Borra (2018) agree to this point when they say that 'none of these [other formats] have (yet) proved as popular as the network diagram in relation to discussions of quali-quantitative methods' (Moats and Borra 2018: 4).

[5] http://www.mappingcontroversies.net/.

sequence analysis (Abbott 1995; Gauthier et al. 2010) to extend the tradition of the Edinburgh School in the Social Shaping of Technology (Williams and Edge 1996) to the new realm of Digital Research (Dutton 2013). It follows in the footsteps of scholars that, after having studied the evolution of careers of IT professionals for over two decades (Williams and Procter 1998), called for extending qualitative sociological methods to new frontiers (Pollock and Williams 2008; Hyysalo 2010), including those offered by new forms of data.

Science Technology Studies and Social Data Science

While there is an emerging body of critical, social-theory-informed digital social research in Critical Geography (Kitchin 2014; Kitchin and McArdle 2016), Critical Data Studies (e.g. Dalton et al. 2016) and Critical Quantification (Bergmann 2016), this book engages social data science from the vantage point of Science and Technology Studies.[6] STS has historically dealt with its objects of study—socio-technological systems and research fields—with a highly critical eye, and thus offers a number of tools for the purpose of this book. Additionally, a vast number of STS studies already exist analysing the role of scientific fields where Big Data computation has been the dominant paradigm such as finance (MacKenzie 2018) or physics (Knorr Cetina 1999).

There are, however, few direct precedents for an STS-inspired study of the scientific process of social data science. This is because such a study requires investigating qualitative social science (in this case ethnography) with the same scrutiny as the investigation of the other contributing sciences (namely statistics and computer science). As noted by Greiffenhagen et al. (2011), while there is a large body of studies on the working practices of statisticians (Hacking 1975; Fine 1973; MacKenzie 1981) and computer scientists (Suchman 1987[2007]; Kunda 1991), the equivalent of laboratory work in the social science is far smaller.[7]

Even when research programmes take impetus from the paucity of empirical studies of how interdisciplinarity unfolds in practice, the interdisciplinary fields considered have been those between natural science and engineering, on the one hand, and the social science, humanities and arts, on the

---

[6] Another important frame of reference for the intellectual agenda described in this book is the Digital STS project initiated by David Ribes at the 4S conference in Cleveland in 2011 and recently summarised in the 'DigitalSTS: A Field Guide for Science & Technology Studies' (Vertesi and Ribes 2019).

[7] Exceptions include Becker (1986) and Clifford and Marcus (1986).

other (Barry and Born 2014; Roepstorff and Frith 2012). As in the case of Greiffenhagen et al.'s (2011) influential study, this book will conduct a symmetrical investigation of the process of scientific work in the three disciplines that constitute social data science including its more qualitative component, hopefully helping more researchers appreciate the strange alliance of phenomenological sociology, computer science and social statistics.

PHENOMENOLOGICAL SOCIOLOGY, COMPUTER SCIENCE AND SOCIAL STATISTICS

Social Data Science is a multilogue which involves a variety of approaches and sub-disciplines. This book will mainly deal with two aspects: (i) the history of ethnography's collaboration with computer science (Fele 2009; Suchman 2007[1987]; Dourish 2001) and (ii) the debate between social phenomenology and social statistics (Snow 1959; Hindess 1973; Collins 1984; Mahoney and Goertz 2006), vis-à-vis the use of new and emerging forms of data.[8]

Its involvement in data science marks a new chapter of ethnography's rapport with computer science: a chapter in which the 'locus of legitimate interpretation' (Bartlett et al. 2018) of social phenomena is more diffuse. Claims from other applied sciences have been recently treated as capable of colonising social research (McFarland et al. 2016: 32). The received view is that the involvement of ethnography with applied research reveals an agonistic–antagonistic mode of interdisciplinarity (Barry and Born 2014: 12). Ethnography antagonises instrumental expectations by the industry and the narrowly mechanical understanding of the social by collaborating computer science disciplines. The association of ethnography and computer science also antagonises existing sociological approaches (Randall et al. 2005). For reasons explained in the section above, the least practised antagonism generated by its interdisciplinary involvement in data science is the one that reflectively addresses ethnography's own presuppositions. This book will make a new proposal in this sense, discussing how social data science brings its own modifications to what empirical sociology currently conceives as workable sociological problems.

The second aspect concerns the well-known fragmentation within the discipline. Social data science puts a new spin on the well-rehearsed

---

[8] Metzler (2016) survey of Big Data research in the social sciences found that, out of 3077 respondents involved in Big Data research, just over half (1690) had most recently used administrative data, 927 used social media data, and 697 used commercial/propriety data.

argument of the split of sociology in two antagonistic wings (the quantitative and the non-quantitative). While the condemnation of the counterpart used to take place *in absentia*, the 'found' character of most big social data brings statistical sociologists to engage with larger disciplinary concerns (Metzler 2016) arguably allowing the two opposing traditions to converge (Latour et al. 2012). Important efforts have recently been made to reflect on the purported divide between computer and social science (Di Maggio 2015). These have addressed statistical social science in particular, recognising that differences in perspective to computer science do not represent any significant disciplinary divide. However, so far machine learning benefits and limitations have been discussed mainly in relation to classical statistical techniques (Molina and Garip 2019). This book will put an accent on the case where the 'two cultures' (Snow 1959) are phenomenological sociology and statistical social science (Collins 1984). Based on fortunate circumstances of collaboration between these complementary positions, I will argue that more than a new name the discipline needs novel case studies that can examine and illustrate how these positions interrelate in concrete settings.

Among the many names that the contemporary form of interdiscipline including qualitative, quantitative social science and computer science can take, I chose for this book the term Social Data Science.[9] The term has been adopted at the Alan Turing Institute (ATI), the UK Centre for Data Science, by the Special Interest Group that I contributed to establish in 2016 with Rob Procter as part of my ATI Faculty Fellowship to shape the Institute's research agenda this area. After reading this book, I hope social scientists who share an ethnography background should feel more confident that their expertise can help calibrate new data science methods. They will hopefully also appreciate the modifications that data science can potentially bring to the received understandings of social phenomena.

FROM ETHNOMETHODOLOGY TO INFRASTRUCTURE STUDIES AND SOCIAL DATA SCIENCE

As a continued effort to describe how data science tools and methods can interact with social science, in this book I will try to describe the continuity between the pre- and the post-data science stages of my research programme.

---

[9] With other candidate names being 'computational social science', 'digital social research', 'big data social science' and 'digital sociology'.

The book reconnects three stages of my research: (i) my early work at the University of Trento with Giolo Fele, taking ethnomethodological concepts to analyse the digital workplace; (ii) my experience as a consultant for the Italian Government's digital agenda and related work conducted for the UKRI CREATe Centre with Robin Williams and Hung The Nguyen; and (iii) my most recent project developed as part of my Alan Turing Faculty Fellowship with Beatrix Alex, Gil Viry and Duncan Chapple. The latter consists in applying text mining and sequence analysis to professional networking data to understand the evolution of careers and expertise in the IT sector. Along the way, the book also describes the circumstances of an evolving research agenda that progressively steps out of the micro-empirical comfort zone to account for larger-scale phenomena. It explores a set of methods and theoretical constructs including situational analysis (Clarke 2005) and linked ecological approaches (Abbott 2005). It suggests these are methodologically reflexive stepping stones for the phenomenologically appraised social researcher towards the more social data science proper.

The journey between the two ends of my programme has been influenced in important ways by concerns emerging from the field of Infrastructure Studies, a field that Dutton recognises as an important component of digital research (Dutton 2013: 188). Around the time when I founded with Robin Williams the 'Innovation in Information Infrastructures' bi-annual workshop with the support of the ESRC Digital Social Research (DSR) programme in 2012, technologically mediated situations were becoming too extended to be dealt with through existing social research methods. Infrastructure studies scholars were increasingly addressing how learning, knowing and sense-making were achieved across the situations that characterise global infrastructure that may be far-removed in time and space. This means my original approach to social data science derives from an understanding of how physical changes in the workplace are paralleled by changes in the method of social inquiry (Law and Urry 2004). This way of doing theory in continuous relation to the distinctive ways of thinking during fieldwork is what Marcus calls the para-site (Hadi and Marcus 2011). The para-site is a kind of deliberate experimental interruption or 'disruption' in the field research process with the intent of staging a reflexive encounter between research partners.

This type of access arguably helped frame the main argument of this book. Any epistemic strategy—digital or not, academic or commercial—brings with it general theoretical constraints: ontology, theories of structure, causality and a concept of time. Combining them along with a cohesive research agenda brings into the open the blind spots of each approach and can help identify phenomena that can be obscured by singular methods.

The book is divided in three parts. The first part describes how ethnomethodological and phenomenological sociology notions can be developed to account for distributed practices. In this part of the book, I will critically comment on the often-considered, context-specific and seemingly anecdotal legacy of workplace studies.

The second part is dedicated to an exploration of social research methods that are predicted on the basis of providing a more strategic approach to the extendedness of technology-mediated practices. It also offers a comparison between these and data-driven topological network approaches.

The third part discusses how to organically expand the approaches discussed in part one and two towards the digital analysis of processual phenomena, interdisciplinarity and new forms of data. It will show how the machine learning artillery can extend ethnography. To conclude, it will offer a view on why ethnography should be considered as a core discipline in data science.

Substantive Research Interest as Method

The three parts should also be read as the components of a personal intellectual biography, where the substantive interest for the IT sector from system design to the study of business expertise is seen as an integral part of 'method'. Each part is introduced by a note of how the journey towards data science has been indeed facilitated by a continued ethnographic engagement with a data-intensive empirical domain.

In the introduction to Chap. 2, I will talk about how the application in earlier works of ethnomethodological concepts helped acquire a critical perspective on computer science expertise. The observation of IT experts' quotidian work helped contextualise the received vision of computer science as 'mathematical problem solving'. Fieldwork revealed that precision and verifiability are used in practice as much as tactics including vagueness (Campagnolo and Fele 2010) and ambiguity (Irons 1998). What is discussed in Chap. 2 is crucial also to understand the view expressed in Chap. 4 on how the distal view of statistical science on social phenomena can complement ethnography's proximal view.

In the introduction to Chap. 3, I will account for my experience as a consultant for the Italian Government digital agenda. In that chapter,

I will talk about what IT projects failure taught me in terms of being able to see technology as fundamentally co-dependent and relationally constituted between computational and discursive practices. What I say in Chap. 3 helps understand the ethnographic view on algorithmic knowledge that I am going to propose in Chap. 5.

In the introduction to Chap. 4, I will describe how this earlier work arguably helped later in my research to build the necessary 'contributory expertise' (Collins and Evans 2007) to engage on a level playing field with colleagues proposing more quantifiable, computational approaches. Long-standing engagement with the IT sector as an empirical field makes a methodological point also at a more general level. Phenomenological, ecological or social data science approaches I alternated in my research were geared towards an understanding of the expertise of system engineers, sales managers, management consultants, innovators and analysts in the IT sector. In a time where data science leads sociological imagination to be shaped by whatever new phenomenon can be studied by transactional data and machine learning methods (e.g. recommendation systems, influence, rumour, virality, fake news...), it is a reminder that as empirical social scientists we base our knowledge claims on expertise deriving from long-standing engagement with a particular field of practice (the study of IT professions in my case) as well as on (or, more appropriately, as part of) mastery of a particular method. In the introduction to Chap. 4, I will discuss the role of continued engagement with a substantive area of application in my work with text mining and sequence analysis collaborators.

The Structure of the Chapters

The core of Chap. 2 is dedicated to discuss how the socially/temporally bounded framing seen in interactionist perspectives has been perceived as particularly ill-equipped to get to grips with phenomena that are instantiated at multiple sites and evolve over time (Monteiro et al. 2013). Preoccupation regarding a blind spot left by 'practice-based research in terms of accounting for the relationship between instances of situated use (i.e., work practices) that are separated in space and/or time' (Vaast and Walsham 2009: 540) will be discussed together with the concern that integrated systems such as large-scale packaged enterprise systems make issues of situated forms of action increasingly intractable (Kallinikos 2004). Instead of looking for alternative theoretical frameworks to interactionism, the chapter will demonstrate that the debate misses the more nuanced treatment of temporality in ethnomethodology (Garfinkel 1967) and phenomenological sociology, preventing appreciation of the vernacular ways through which business actors conceptualise, refer to and make use of the 'trans-local'. In accordance with scholars who suggest that locating ethnomethodology at the 'micro' end of the sociological spectrum is a misleading characterisation (Bittner 1965; Coulter 1996), the chapter suggests that concepts can be found in this tradition to account for how members make sense of what is beyond their reach and help liberate the analytical potential of situated studies from the legacy of a focus on short durations and small spaces. The chapter will conclude with a discussion on the formative effects of digital practice on social science theorizing.

Chapter 3 discusses the application of analogue mapping methods such as situational analysis (Clarke 2005) and the linked ecologies approach (Abbott 2005) whose key point is to also offer ways to represent connections among apparently unrelated organisational settings or even sectors and support the understanding of changes in discourse and actors over time. With these approaches, there is no assumption of any territorial limitation to what 'local' means. Through extensive and strategic data collection across sites and over time, the research projects discussed in this chapter connect distant situations and make them, at least temporarily and partially, 'practically one' (Stump 1996). These approaches will be exemplified by their application to two cases: one is the case of the early introduction of packaged enterprise software in the Italian public sector over a ten-year period and across a number of interconnected field sites. The other follows the development of a digital infrastructure for intellectual property trading in the UK. By comparing analogue maps with digital network graphs, I try to account for an apparent lack of ambition in the digital social research agenda. I argue that failure to challenge existing sociological concepts (or reliance on sociological theories of the time past) is actually one of the reasons for the methodological naivety of many current digital research approaches.

In Chap. 4, I will discuss why and how a programme of research mainly based on the above-mentioned methods has been extended to encompass computational and statistical methods. This chapter offers a new approach to discussing social data science, aimed at overcoming the "service" relationship whereby the social scientist can be seen as a 'problem owner', whose conjectures can be validated by the data scientist through a mixture of assumption-free computational and statistical models.

The default problem for ethnographic engagements with data science is the opacity of computational practices (Jaton 2017). Algorithms and statistical models present themselves to the researcher as black boxes (Seaver

2017). Fortunate experiences of collaboration and a positioning in the team as the holder of "ground truth" meant that in the project discussed in Chap. 4, I had the chance to directly affect the conduct of the computational and statistical operations. Following Knorr-Cetina's work on epistemic cultures (Knorr Cetina 1999), in this chapter I will describe how the ethnographer's standpoint can influence (i) the human collaboration as well as the selection of the digital platform producing data; (ii) the objects of scientific research, and the reconfigurations through which the object gradually becomes amenable to the analysis; and (iii) the object–relation regime, which in the project under study means the interaction between the object of research (i.e. careers) and methodologies (text mining and sequence analysis).

The findings from Chap. 4 indicate that there is a two-way interaction between social and data science. Responding to the call from its own proponents that data science is at a point where a study of its own ways of working is desperately needed, in the concluding chapter (Chap. 5) I will devise a strategy for ethnographers to gain a practical foothold in the area of data science they want to analyse. I will address the benefits for data science when social scientists directly contribute to as well as observe data science projects. More precisely, I will discuss how ethnographic accounts can help legitimate data science contribution more broadly and beyond the initial promises linked to successful but localised applications. Starting from a review of current literature on the socio-materiality of algorithms, the chapter will identify the architectural aspect of machine learning as the next focus of social research. Revisiting the debate between ethnomethodology inspired ethnography and symbolic AI as well as the sociology of scientific knowledge argument on expert systems, I will ask what the contact points are between phenomenological sociology and the resurgent connectionism. Adopting a deflationary sensibility, I will critically address the claims of naturalness that surround contemporary deep machine learning approaches. Revisiting the ethnomethodological notion of accountability (Garfinkel 1967: 1) and how it has been mobilised by the technomethodology programme (Dourish and Button 1998) and more recently by social studies of algorithms (Neyland 2016) I will identify the explainable AI (xAI) debate as a candidate common ground for current interdisciplinary conversations. My conclusions are that given the direction machine learning is taking of creating more and more diverse model ensembles, the next big task of social data science is to understand how algorithms interact.

REFERENCES

Abbott, A. (1992). From Causes to Events: Notes of Narrative Positivism. *Sociological Methods & Research, 20*(4), 428–455.

Abbott, A. (1995). Sequence Analysis: New Methods for Old Ideas. *Annual Review of Sociology, 21*, 93–113.

Abbott, A. (2005). Linked Ecologies: States and Universities as Environments for Professions. *Sociological Theory, 23*(3), 245–274.

Anderson, C. (2008). *The End of Theory: The Data Deluge Makes the Scientific Method Obsolete.* Wired Magazine. Retrieved from 23 June, 2008, from https://www.wired.com/2008/06/pb-theory/.

Atkinson, P. A., & Housley, W. (2003). *Interactionism.* London: SAGE.

Barry, A., & Born, G. (2014). *Interdisciplinarity: Reconfigurations of the Social and Natural Sciences.* London: Routledge.

Bartlett, A., Lewis, J., Reyes-Galindo, L., & Stephens, N. (2018). The Locus of Legitimate Interpretation in Big Data Sciences: Lessons for Computational Social Science from -Omic Biology and High-Energy Physics. *Big Data & Society, 5*(1), 1–15.

Beaulieu, A. (2016). Vectors for Fieldwork: Computational Thinking and New Modes of Ethnography. In L. Hjorth, H. Horst, A. Galloway, & G. Bel (Eds.), *In Companion to Digital Ethnography* (pp. 29–39). London: Routledge.

Becker, H. (1986). *Writing for Social Scientists: How to Start and Finish Your Thesis, Book, or Article.* Chicago: University of Chicago Press.

Bergmann, L. (2016). Toward Speculative Data: "Geographic Information" for Situated Knowledges, Vibrant Matter, and Relational Spaces. *Society and Space, 34*(6), 971–989.

Bittner, E. (1965). The Concept of Organization. *Social Research, 32*(3), 239–255.

Bouillier, D. (2018). Médialab Stories: How to Align Actor Network Theory and Digital Methods. *Big Data & Society, 5*(2), 1–13.

Burnap, P., Rana, O., Williams, M., et al. (2014). COSMOS: Towards an Integrated and Scalable Service for Analyzing Social Media on Demand. *International Journal of Parallel, Emergent and Distributed Systems (IJPEDS), 30*(2), 80–100.

Campagnolo, G. M., & Fele, G. (2010). From Specifications to Specific Vagueness: How Enterprise Software Mediates Engineering Relations. *Engineering Studies, 2*(3), 221–243.

Clarke, A. E. (2005). *Situational Analysis: Grounded Theory After the Postmodern Turn.* Thousand Oaks, CA: SAGE.

Clifford, J., & Marcus, G. E. (Eds.). (1986). *Writing Culture: The Poetics and Politics of Ethnography.* Berkeley: University of California Press.

Collins, R. (1984). Statistics Versus Words. *Sociological Theory, 2*, 329–362.

Collins, R. (1994). Why the Social Sciences Won't Become High-Consensus, Rapid-Discovery Science. *Sociological Forum, 9*(2), 155–177.

Collins, H. M., & Evans, R. (2007). *Rethinking Expertise*. Chicago, IL: University of Chicago Press.

Coulter, J. (1996). Human Practices and the Observability of the 'Macrosocial'. *Zeitschrift für Soziologie, 25*, 337–345.

Dalton, C., Taylor, L., & Thatcher, J. (2016). *Critical Data Studies: A Dialog on Data and Space*. SSRN. Retrieved from https://ssrn.com/abstract=2761166.

Di Maggio, P. (2015). Adapting Computational Text Analysis to Social Science (and Vice Versa). *Big Data & Society, 2*(2), 1–5.

Dourish, P. (2001). *Where the Action Is: The Foundations of Embodied Interaction*. Cambridge, MA: MIT Press.

Dourish, P., & Button, G. (1998). On "Technomethodology": Foundational Relationships Between Ethnomethodology and System Design. *Human-Computer Interaction, 13*(4), 395–432.

Dutton, H. W. (2013). The Social Shaping of Digital Research. *International Journal of Social Research Methodology, 16*(3), 177–195.

Edwards, A., Housley, W., Williams, M., Sloan, L., & Williams, M. (2013). Digital Social Research, Social Media and the Sociological Imagination: Surrogacy, Augmentation and Re-orientation. *International Journal of Social Research Methodology, 24*, 313–343.

Fele, G. (2009). Why is Information System Design Interested in Ethnography? Sketches of an Ongoing Story. *Ethnografia e Ricerca Qualitativa, 1*, 1–38.

Fine, T. (1973). *Theories of Probability: An Examination of Foundations*. New York: Academic Press.

Garfinkel, H. (1967). *Studies in Ethnomethodology*. Englewood Cliffs, NJ: Prentice-Hall.

Gauthier, J. A., Widmer, E. D., Bucher, P., & Notredame, C. (2010). Multi-channel Sequence Analysis Applied to Social Science Data. *Sociological Methodology, 40*, 1–38.

Gouldner, A. W. (1970). *The Coming Crisis of Western Sociology*. New York: Basic Books.

Greiffenhagen, C., Mair, M., & Sharrock, W. (2011). From Methodology to Methodography: A Study of Qualitative and Quantitative Reasoning in Practice. *Methodological Innovations Online, 6*(3), 93–107.

Hacking, I. (1975). *The Emergence of Probability*. Cambridge: Cambridge University Press.

Hadi, D., & Marcus, G. E. (2011). In the Green Room: An Experiment in Ethnographic Method at the WTO. *PoLAR, 34*(1), 51–76.

Halfpenny, P., & Procter, R. (2015). *Innovations in Digital Research Methods*. London: Sage.

Hindess, B. (1973). *The Use of Official Statistics in Sociology: A Critique of Positivism and Ethnomethodology*. London: Macmillan.

Housley, W., & Smith, R. J. (2017). Interactionism and Digital Society. *Qualitative Research 17*(2), 187–201.

Hyysalo, S. (2010). *Health Technology Development and Use: From Practice-Bound Imagination to Evolving Impacts*. London: Taylor & Francis.

Irons, R. L. (1998). Organizational and Technical Communication: Terminological Ambiguity in Representing Work. *Management Communication Quarterly, 12*(1), 42–71.

Jaton, F. (2017). We Get the Algorithms of Our Ground Truths: Designing Referential Databases in Digital Image Processing. *Social Studies of Science, 47*(6), 811–840.

Kallinikos, J. (2004). Farewell to Constructivism: Technology and Context-embedded Action. In C. Avgerou, C. Ciborra, & F. Land (Eds.), The Social Study of Information and Communication Technology: Innovation, Actors, and Contexts. Oxford: Oxford University Press.

Kitchin, R. (2014). Big Data, New Epistemologies and Paradigm Shifts. *Big Data & Society, 1*(1), 1–12.

Kitchin, R., & McArdle, G. (2016). What Makes Big Data, Big Data? Exploring the Ontological Characteristics of 26 Datasets. *Big Data & Society, 3*(1). https://doi.org/10.1177/2053951716631130.

Knorr Cetina, K. (1999). *Epistemic Cultures: How the Sciences Make Knowledge*. Cambridge, MA: Harvard University Press.

Kunda, G. (1991). *Engineering Culture: Control and Commitment in a High-Tech Corporation*. Philadelphia: Temple University Press.

Latour, B., Jensen, P., Venturini, T., Grauwin, S., & Bouillier, D. (2012). 'The Whole Is Always Smaller Than Its Parts'—A Digital Test of Gabriel Tardes' Monads. *The British Journal of Sociology, 63*(4), 590–615.

Law, J., & Urry, J. (2004). Enacting the Social. *Economy and Society, 33*(3), 390–410.

Lazer, D., & Radford, J. (2017). Data Ex Machina: Introduction to Big Data. *Annual Review of Sociology, 43*, 19–39.

Lupton, D. (2014). *Digital Sociology*. London and New York: Routledge.

MacKenzie, D. (1981). *Statistics in Britain, 1865–1930: The Social Construction of Scientific Knowledge*. Edinburgh: Edinburgh University Press.

MacKenzie, D. (2018). 'Making', 'Taking' and the Material Political Economy of Algorithmic Trading. *Economy and Society, 47*(4), 501–523.

Mahoney, J., & Goertz, G. (2006). A Tale of Two Cultures: Contrasting Quantitative and Qualitative Research. *Political Analysis, 14*(3), 33–53.

Marres, N. (2017). *Digital Sociology*. Cambridge: Polity Press.

Marres, N., & Moats, D. (2015). *Mapping Controversies with Social Media: The Case for Symmetry*. SSRN. Retrieved from https://ssrn.com/abstract=2567929 or https://doi.org/10.2139/ssrn.2567929.

McFarland, A. D., Lewis, K., & Goldberg, A. (2016). Sociology in the Era of Big Data: The Ascent of Forensic Social Science. The *American Sociologist, 47*, 12–35.

Metzler, K. (2016). *The Big Data Rich and the Big Data Poor: The New Digital Divide Raises Questions About Future Academic Research*. The Impact Blog, London School of Economics and Political Science. Retrieved from http://blogs.lse.ac.uk/impactofsocialsciences/2016/11/22/the-big-data-rich-and-the-big-data-poor-the-new-digital-divide-raises-questions-about-future-academic-research/.

Mills, C. W. (1959). *The Sociological Imagination*. Oxford: Oxford University Press.

Moats, D., & Borra, E. (2018). Quali-Quantitative Methods Beyond Networks: Studying Information Diffusion on Twitter with the Modulation Sequencer. *Data & Society, 5*(1), 1–17.

Molina, M., & Garip, F. (2019). Machine Learning for Sociology. *Annual Review Sociology, 45*, 27–45.

Monteiro, E., Pollock, N., Hanseth, O., & Williams, R. (2013). From Artefacts to Infrastructure. *Computer Supported Cooperative Work, 22*(4–6), 575–607. (CSCW).

Neyland, D. (2016). Bearing Account-able Witness to the Ethical Algorithmic System. *Science, Technology, & Human Values, 41*(1), 50–76.

Orton-Johnson, K., & Prior, N. (Eds.). (2013). *Digital Sociology: Critical Perspectives*. London: Palgrave Macmillan.

Pentland, B. (2013). Desperately Seeking Structures: Grammars of Action in Information Systems Research. *The DATA BASE for Advances in Information Systems, 44*(2), 7–18.

Pentland, B.T., Hærem, T., & Hillison, D. (2011). The (N)Ever-Changing World: Stability and Change in Organizational Routines, *Organization Science, 22*(6), 1369–1383.

Pentland, B. T., Feldman, M. S., Becker, M. C., and Liu, P. (2012). Dynamics of Organizational Routines: A Generative Model, *Journal of Management Studies, 49*(8), 1484–1508.

Pollock, N., & Williams, R. (2008). *Software and Organisations: The Biography of the Enterprise-Wide System or How SAP Conquered the World*. Oxon: Routledge.

Procter, R., Vis, F., & Voss, A. (2013). Reading the Riots on Twitter: Methodological Innovation for the Analysis of Big Data. *International Journal of Social Research Methodology, 16*(3), 197–214.

Randall, D., Harper, R., & Rouncefield, M. (2005). Fieldwork, Ethnography and Design: A Perspective from CSCW. In K. Anderson & T. Lovejoy (Eds.), *EPIC 2005: Ethnographic Praxis in Industry Conference* (pp. 88–99). Seattle, WA and Arlington, VA: Redmond, American Anthropological Association.

Rheinberger, H.-J. (2011). Consistency from the Perspective of an Experimental Systems Approach to the Sciences and Their Epistemic Objects. *Manuscrito, 34*(1), 307–321.

Roepstorff, A., & Frith, C. (2012). Neuroanthropology or Simply Anthropology? Going Experimental as Method, as Object of Study, and as Research Aesthetic. *Anthropological Theory, 12*(1), 101–111.

Rogers, R., & Marres, N. (2000). Landscaping Climate Change: A Mapping Technique for Understanding Science and Technology Debates on the World Wide Web. *Public Understanding of Science, 9*(2), 141–163.

Salganik, M. J. (2018). *Bit by Bit: Social Research in the Digital Age.* Princeton, NJ: Princeton University Press.

Sandvig, C., & Hargittai, E. (2015). How to Think about Digital Research. In E. Hargittai & C. Sandvig (Eds.), *Digital Research Confidential: The Secrets of Studying Behavior Online.* Cambridge, MA: MIT Press.

Savage, M. (2015). Sociology and the Digital Challenge. In P. Halfpenny & R. Procter (Eds.), *Innovations in Digital Research Methods.* London: Sage.

Savage, M., & Burrows, R. (2007). The Coming Crisis of Empirical Sociology. *Sociology, 41*(5), 885–899.

Seaver, N. (2017). Algorithms as Culture: Some Tactics for the Ethnography of Algorithmic Systems. *Big Data & Society, 4*(2), 1–12.

Snow, C. P. (1959). *The Two Cultures and the Scientific Revolution.* Cambridge: Cambridge University Press.

Stump, David, J. 1996. From Epistemology and Metaphysic s to Concrete Connections', in D. Stump and P. Galison (eds), Disunity of Science: Boundaries, Contexts, and Power. Stanford, CA: Stanford University Press, pp. 255–286.

Suchman, L. A. (1987). *Plans and Situated Actions: The Problem of Human-Machine Communication.* Cambridge: Cambridge Press.

Vaast, E., & Walsham, G. (2009). Trans-Situated Learning: Supporting a Network of Practice with an Information Infrastructure. *Information Systems Research, 20*(4), 547–564.

Vedres, B., & Stark, D. (2010). Structural Folds: Generative Disruption in Overlapping Groups. *American Journal of Sociology, 115*(4) :1150–1190.

Veltri, A. G. (2019). *Digital Social Research.* Cambridge: Polity Books.

Venturini, T., Jacomy, M., & Meaner, A. (2017). An Unexpected Journey: A Few Lessons from Sciences Po Médialab's Experience. *Big Data & Society, 4*(2), 1–11.

Vertesi, J., & Ribes, D. (2019). DigitalSTS: A Field Guide for Science & Technology Studies. Princeton University Press.

Williams, R., & Edge, D. (1996). The Social Shaping of Technology. *Research Policy Vol., 25*(1996), 856–899.

Williams, R., & Procter, R. (1998). Trading Places: A Case Study of the Formation and Deployment of Computing Expertise. In R. Williams et al. (Eds.), *Exploring Expertise* (pp. 197–222). Basingstoke: Macmillan. Chap. 13.

CHAPTER 2

Phenomenological Extensions

Abstract This chapter shows how continued ethnographic engagement with data-intensive domains can help acquiring a critical perspective on data science expertise. It reviews the early success of symbolic interactionism and ethnomethodology in human–computer interaction (HCI), artificial intelligence (AI) and computer-supported cooperative work (CSCW). It then discusses how with the ever-increasing extendedness of technologically mediated situations, the viability of situated studies of technology and work came under increased scrutiny for not covering extended settings. By referring to Garfinkel's notion of 'specific vagueness' and to Schutz's notion of 'appresentation', it presents aspects of phenomenological sociology that remain valid in the context of digital practices.

Keywords Phenomenological sociology • Ethnomethodology • Vagueness • Appresentation

In this chapter I present my early work taking ethnomethodological concepts to analyse the digital workplace as the first step of a journey towards social data science. In line with the recurring theme of the book to conceive personal intellectual biographies as an integral part of method, that is, 'the researcher is an instrument', I describe how the continued ethnographic engagement with the IT sector as a data-intensive domain helped

G. M. Campagnolo, *Social Data Science Xennials*, https://doi.org/10.1007/978-3-030-60358-8_2

acquiring a critical perspective on data science expertise. In particular, I show how the observation of IT experts' quotidian work helped contextualise the received vision of computer science. Rather than just 'mathematical problem solving', it revealed that precision and verifiability are used in practice as much as tactics including vagueness (Campagnolo and Fele 2010) and ambiguity (Irons 1998).

Subsequently, I describe how the viability of situated studies of technology and work came under increased scrutiny for not covering extended settings in the context of the development of an alternative approach called biography of artefacts (Pollock and Williams 2009).

In this chapter, I demonstrate how ethnomethodological and phenomenological sociology notions can be developed to account for distributed practices. I critically comment to the often-considered context-specific and seemingly anecdotal legacy of workplace studies and discuss how the socially/temporally bounded framing seen in interactionist perspectives has been perceived as particularly ill-equipped to get to grips with phenomena that are instantiated at multiple sites and evolve over time (Monteiro et al. 2013). Instead of looking for alternative theoretical frameworks to interactionism, I show that the debate misses the more nuanced treatment of temporality in ethnomethodology (Garfinkel 1967) and phenomenological sociology, preventing appreciation of the vernacular ways through which actors conceptualize, refer to and make use of the 'trans-local'.[1]

As well as appreciating what aspects of the phenomenological approach remain valid, in conclusion I will begin to address a topic that dominates this book, that is, how the digital challenges ethnographic work. In this chapter, the tension between digital and analogue is explored at a conceptual level. Going back to its roots in ethnomethodology, I will ask what it takes for ethnography to attend digital practices. While in Chap. 4 the same topic will be explored at a more methodological level in a case where digital methods and ethnographic methods co-occur in the same study.

---

[1] In this respect, Abbott (1992) attributes to empirical literatures such as symbolic interactionism and ethnomethodology an ability to acknowledge the importance of process that is ignored by the main empirical traditions of sociology.

VAGUENESS

The study I want to begin with to demonstrate the formative effect of the research object (the IT sector) on my own research approach is an ethnography of software demonstrations given by members of the product development and sales division of a system engineering company to various industrial design engineers in the transition of system engineering services to the international market of enterprise architecture (Campagnolo and Fele 2010).

I was conducting observations on demonstrations given by a system engineering company's sales people to different sorts of design engineers—namely planning engineers, electronic engineers and production engineers. The purpose of these demonstrations was to discuss the potential uses and technological possibilities of a software visual modeling tool being proposed within the project for the purpose of visualising industrial manufacturing processes.

At the time, I was interested in an ever-growing tendency in the contemporary enterprise architecture software market for vendors to package ready-to-install products, contrary to the previous tendency whereby organisations themselves developed their own software for their needs. This meant that system engineering software tools were increasingly becoming generic software tools designed to cover the entire set of problems that a wide range of user organisations may face. In many ways, these enterprise architecture software platforms are open toolboxes: user organisations have to decide for themselves what methodology to use and what modeling to do, a tasks which only mature organisations can undertake on their own. With my study, I wanted to observe the implications of these major changes in the enterprise architecture software market for concrete system engineering practices. System engineering companies increasingly need to build 'partnerships' and 'competence centers' on the user side. They must be able to customise generic tools and to resolve minor adaptation problems locally. In other words, rather than being seen as 'mathematical problem solving' in which 'dominant practices were framing service as solving problems given by others and the agents involved were coming to expect both themselves and others to be either right or wrong' (Downey 2009: 61) when mediated by generic enterprise architecture software, engineering problem solving becomes a distributed practice. This involves numerous agents with different degrees of responsibility for the realisation of technological possibilities. As a consequence, rather than solving problems, the concern of sales system engineers is to make problems vague enough so that they can respond to the enterprise architecture market's logics and specific enough to be used for the purpose at hand in a particular industrial context.

With many hours of audio–video recording, I captured negotiations between system engineers and design engineers on a graphical software modeling tool which focused on (i) the amount of its functions, (ii) the range of its functions and (iii) their degree of integration. A recurring evidence was that being 'specifically vague' about the potential uses and technological possibilities of the software graphical modeling tool was the key issue at stake in the negotiation process between system engineers and design engineers. Each also demonstrates that system engineers are vague about different aspects of software functionalities according to the specific design engineering domain they encounter.

Contrary to expectations concerning engineering when it is viewed as a matter of logic, the establishment of a common ground of reference between system engineers and design engineers of various kinds did not seem to require system engineers to solve design engineers' problems, but rather to make them sufficiently vague to be matched by the generic answer provided by the system.

The main theoretical reference of this study has been the seminal work of the founder of ethnomethodology, Harold Garfinkel (1967), who itemises the 'specific vagueness of reference' among the sanctioned properties of common discourse that persons use to permit each other to understand what they are really talking about (Garfinkel 1967: 41). Being 'specifically vague' allows for the fundamental 'understanding work' that enables communication between people to develop and thereby generate a shared horizon of interpretations and expressions among interactants. According to Garfinkel, events that are talked about are 'specifically vague' in that:

> Not only do they not frame a clearly restricted set of possible determinations but the depicted events include as their essentially intended and sanctioned features an accompanying 'fringe' of determinations that are open with respect to internal relationships, relationships to other events, and relationships to retrospective and prospective possibilities. (Garfinkel 1967: 41)

Garfinkel uses ingenious experiments to show that the meaningful orderliness of a social situation is hampered if the interactants seek excessively to be 'more precise', 'more accurate', or 'more detailed'. Contrary to our common understanding of communication processes as a matter of logic, whenever an agreement is reached between two parties, the events that are talked about are never fully specified. That is to say, it is assumed that the two parties, in order to reach mutual understanding and to agree upon a common reference, do not need to specify in advance any possible meaning, use, function, or interpretation of what is being talked about. From our point of view, the notion that 'being specific' will assure joint agreement and shared understanding is just as valid as the reverse ('being vague').

My approach to vagueness had resemblances with findings from other naturalistic studies of engineering workplace interactions that also suggested interpretation as constitutive of engineering negotiations. In these studies, incomplete and unstable specifications were seen as constitutive of system development. For example, Suchman et al. (2002) describe how basic assumptions regarding the status of 'user needs' can be reversed. They point out that for most advocates the role of engineers is to uncover user needs that are already existing but somehow latent and unrecognized by prospective users. Suchman et al. argue instead that engineering practice 'simultaneously recover and invent work requirements and technological possibilities, that each make sense in relation to the other' (2002: 166). Another similar concept that has been applied to the study of technical specifications is that of 'strategic ambiguity' (Eisenberg 1984: 230). Irons (1998) used the concept of strategic ambiguity to study the workplace task of representing work through technical communication. As noted by Eisenberg (1984: 239) he found that 'ambiguous communication is not a kind of fudging, but rather a rational method used by communicators to orient toward multiple goals'.

These early studies that revealed the quotidian aspects in the practice of designing complex technological architectures helped my determination in pursuing a research agenda that blurred the distinction between social sciences and computer science.[2]

---

[2] This research agenda has been deeply inspired also by Claudio Ciborra (2000), a leading advocate of the separation of informatics from the mathematical and formal sciences, and its inclusion among the social and human sciences and by Paul Dourish (2001) idea of 'technomethodology' whereby ethnomethodology could prove the foundational concept for the design of information architectures. More on technomethodology in Chap. 5.

APPRESENTATION

After the early success of symbolic interactionism and ethnomethodology in the field of studies on human–computer interaction (HCI), artificial intelligence (AI) and computer-supported cooperative work (CSCW), with the ever-increasing extendedness of technologically mediated situations, the viability of approaches that dealt with sense-making (i.e. situated studies of technology and work) came under increased scrutiny for not covering extended settings. These were key studies in revealing that workplace technologies cannot be understood as particular finite and thus knowable technical systems but are instead accomplishments of a complex dialectic relationship between discursive and material practice (Hutchins 1996; Lave 1988; Orr 1996; Suchman 1987). However, more recently in the field of Infrastructure Studies there has been discussion on how existing modes of research inspired by interactionism frame the analysis (Ciborra 2006; Pollock and Williams 2009: 80). The argument is that most studies influenced by interactionist perspectives are framed, somewhat unreflexively, by particular socially/temporally bounded locales. The socially/temporally bounded framing seen in interactionist perspectives seems, according to these authors, particularly ill-equipped to get to grips with complex technologies that are instantiated at multiple sites and evolve over time (Monteiro et al. 2013). This argument resounds with a similar preoccupation voiced by Vaast and Walsham (2009), who recognise that there is a blind spot left by 'practice-based research in terms of accounting for the relationship between instances of situated use (i.e., work practices) that are separated in space and/or time' (Vaast and Walsham 2009: 540). Similarly, Kallinikos (2004) focuses on how large-scale corporate integrated systems make issues of situated forms of action increasingly intractable.

This debate in Infrastructure Studies anticipated some of the key issues in the discussion that took place few years later in relation with the UKRC supported programme in Digital Social Research (Dutton 2013), namely that innovation in social research methods meant understanding of how physical changes in the workplace are paralleled by changes in the method of social inquiry (Law and Urry 2004).

As a co-organiser of the 'Innovation in Information Infrastructures' bi-annual workshop with the support of the ESRC Digital Social Research (DSR) programme in 2012, my answer at the time has been to demonstrate that the debate missed the more nuanced treatment of temporality in ethnomethodology (Garfinkel 1967) and phenomenological sociology, preventing appreciation of the vernacular ways through which actors in technology-mediated situations conceptualize, refer to and make use of the 'trans-local' (Campagnolo et al. 2015).

One clear case is the concept of 'appresentation'. Schutz and Luckmann (1989: 2, 131–132) use the term 'appresentation' to refer to the process of making available to participants 'what lies spatially and temporally beyond their reach'. Schutz and Luckman inherit this concept from Husserl (1960, 1931), who surmises that 'the strictly seen front of a physical thing always and necessarily appresents a rear aspect and prescribes for it a more or less determinate content' (Husserl 1960, 1931: 109). For Husserl, appresentation is the experience of seeing 'more' than concretely presented perspectival details: seeing that a house has an interior and a back when viewing it from the front. What is not concretely 'present' is implicated along with what is immediately apparent and, reciprocally, it determines the sense of what is seen (a house with an interior, etc.). Schutz further developed the idea of applying to social interaction 'appresentational pairing' (Schutz 1971: 248). This is a way to conceptualise the reciprocal 'perception of the other': the appreciation by a recipient of what a speaker intends with an overt expression, together with the speaker's inner orientation to the recipient's (possible) understanding. As Schutz says: 'we directly perceive the moving body of another person, and we necessarily pair it with our general notion of an alter ego' (Schutz 1971: 321). In phenomenological writing, appresentation is different from representation. Representation applies to the perception of a phenomenon that has been at least once an object of our consciousness. While existing research on extended situations refers to how participants perceive phenomena that have been at least once an object of their consciousness, appresentation refers to something that we always perceive as being present (e.g. the planet Earth in its entirety) with no original experience of it (Rodemeyer 2006: 143).

Appresentation points to how members' sense-making resources should be understood in a context of reciprocal interaction, unfolding in time rather than a particular moment (or locale) of anticipation.

By widening the old phenomenological concept of 'appresentation' I wanted to contrast the view that interactionist perspectives necessarily conceive of sense-making as a somewhat closed, self-compensating, self-terminating circuit of interdependent action, always circumscribed by the small space of the 'workplace' (see Henke 1999).

My point is that appresentation can be seen as an emerging social skill whereby the global IT industry (suppliers and buyers) makes sense of usage sites that are beyond what a vendor can reach for direct information. The empirical account at the time was the appresentation work of global Enterprise Resource Planning (ERP) vendors as well as their clients, who are seen as active participants in shaping enterprise technology. Given the large investment required to procure these systems, buyers cannot afford to learn about the ERP product by trial and error (Tingling and Parent 2004). It is also extremely hard to assess the properties of a packaged enterprise system as these cannot be readily discerned by direct inspection and are only evident when implemented and used by an organisation. As Smith (2009) points out, the audience tends not to agree on what was seen in ERP demonstrations. Vendors and their clients seem therefore to be engaged in a more nuanced activity than just comparing dispersed user sites and technological practices.

In the course of my research, I have been able to study how three global vendors attempt to enrol key users in selling their products in a new domain and often in a new geographical area. One study is of a mid-range enterprise software company (referred to as Metis) with offices in Scandinavia and the US, which at the time of our study began to address the European automotive market. Another study involves one of the world's largest software vendors, which was expanding sales to the Public Sector. In this case, we studied how the vendor worked with a group of early adopters in the Italian public sector, who formed the first public sector user group for that type of software system. The third case is an ERP system (PAMS) developed for Higher Education Institutions, built by a company we call 'Educational Systems'. PAMS was initially designed around the needs of a Scottish University but is now being used by more than 40 other institutions in the UK. At the time of our research the supplier was investigating the potential overseas market.[3]

Based on these three cases, I described how IT vendors and the buyers of their products appresent differences between globally distributed client organisations. Observation of 'appresentation work' in global software development helped reveal the complex social fabric between ERP vendors and their clients. This fabric lets them make sense, and convey some confidence about matters on which they have no direct experience. We

---

[3] Evidence from the third case derives from co-authors Neil Pollock and Robin William fieldwork as is also reported in earlier works, for example Pollock et al. (2007).

focused in particular on how vendors (and users) appresent the global user base. Vendors achieve 'user appresentation' by making users recognisably the same (i) through preparing them to conceive of their requirements in a form that is generic enough to fit the whole user community, (ii) by passing the task of assimilating users over to other users and (iii) by progressively changing the criteria for segmenting users from seniority to future-orientation. Rather than a limitation, it emerged that referring to something that is not known is an essential part of successful sense-making in the extended negotiations that lead to ERP development and use. This applies even when direct inspection is an option. Strategic reference is made to what is not known (i.e. regarding distant and future situations) to make sense of what is available.

In conclusion of that study, I mentioned something that has become central to the subsequent development of my research programme and that arguably sprung my interest for computational methods.

I suggested how our analysis complemented situated learning accounts whereby becoming member of a practice implies socialisation (Nicolini 2011: 613). For example, in providing an account of the 'expansion of practice', Nicolini (2011) argues that 'knowing [has] to do with tuning and becoming absorbed in an ongoing practical regime' (Nicolini 2011: 613). Emphasis on socialisation as a form of learning is also found in the account by Monteiro et al. (2012) of how 'family resemblance' is achieved between time-space separated work practices. Their identification of the three strategies that constitute similarity across settings (differentiation, assembling and punctuation) refers to re-gaining intimate knowledge of each situation and so ultimately to understanding how far past lessons can be re-used.

My analysis addresses how knowing is possible when becoming absorbed in a situation is not an option and it shows that matters that are not known are always and necessarily a component of extended sense-making and how this applies even when direct inspection is an option. Strategic reference is made to what is not known (i.e. regarding distant and future situations) to make sense of what is available.

By reporting the case that even when close interaction is possible, IT vendors preferred not to get involved in specific implementations, I was suggesting that in order to produce knowledge that extends in space-time, knowing how not to interact is as relevant as knowing how to interact. Learning how not to become too embedded in the context of a practice is therefore one neglected social form of extended knowledge production. I argue that in studies of trans-local achievements, more attention should be given to social skills that allow participants to keep (or create) a space-time distance between them and the contingencies in which they operate. In other words, my argument is that we shouldn't be preoccupied that most interactionism-inspired studies of technology are unreflexively framed by particularly socially/temporally bounded locales. What should be more worrying is that, forgetful of the phenomenological legacy, these studies and their results reproduce a patronising view of social interaction as conservative and incapable to engage in forms of adaptation that surpass the horizon of the present, including routinely engaging with the language of 'unknown' (future or large-scale). It is not by chance that, as noted by Fele (2009: 20), ethnographic research has been littered with studies whose conclusions were to leave things as they are.[4]

Elsewhere a great deal of study has been devoted to the role of inter-subjectivity when describing how knowledge developed in a certain context is translated elsewhere (Callon 1986; Ophir and Shapin 1991). However, less attention has been given to the way what is absent contributes to shape the context of knowledge production. Wittgenstein (1953) has been the first to raise the question of the problematic move from case to case. 'What can take us beyond the mere perception of a phenomenon' he asked. How can we know how to apply certain categories that have been elaborated on the basis of a finite number of cases to the next case? Wittgenstein answered the question by saying that we move from case to case by seeing them as similar to past cases. The term Wittgenstein uses is 'analogy' (Wittgenstein 1978: 122–126). We know less about how the next case is used to define how we make sense of the present. This we maintain is the one level of inter-subjectivity that has been neglected so far in interactionist analyses of technological situations.

---

[4] The most notable of these studies have been those by Suchman (1994) on a law firm, by Sommerville et al. (1993) on air traffic controllers, by Hutchins and Palen (1998) on aircraft cockpits, and by Harper and Carter (1994) on a firm of architects.

The Formative Effects of Digital Practices on Social Research

My study of 'appresentation' proposed the ability of the social researcher to self-distance himself or herself from the practice under investigation as a form of knowing. This statement will be expanded in Chap. 4 when I will talk about unconscious biases of the interactionist approach.

In this concluding section, I want to make the case that the ability to infer and make projections that I found in business experts is now increasingly valid also for other domains of sociological inquiry. As social media data makes ordinary social interaction as data intensive as the cases of expert work discussed in this chapter, the increasingly digital nature of social practices is having a formative effect on how we (can) study them.

My argument on the formative effects of data-intensive practices on social research is inspired by Pollner's discussion on the relationship between ethnomethodology and business (Pollner 2012b). Quoting Anne Rawls (2002), Pollner recounts how Garfinkel enrolled in a course called 'Theory of Accounts', which covered the techniques of bookkeeping and accounting. From that course he came 'to understand that even in setting up an accounting sheet, he was theorizing the various categories into which the numbers would be placed' (Rawls 2002: 10).

Before addressing in the next chapters how alternative qualitative, quali-quantitative and 'commercial sociology' methods can be used to study digitally mediated practices, in the concluding section of this chapter I want to extend the "formative effects argument" from IT business to social interaction more widely.

In this chapter, I argued that in global IT business the issue of coordination in extended situations becomes a concrete issue: the issue of what connects distant situations and where they connect. The increasing availability of social data in the form of digital traces similarly puts under the spotlight the question of what connects distant practices and makes them, at least temporarily and partially, 'practically one' (Stump 1996). Complete digital data sets at the level of the population such as those discussed for example in Chap. 4 put the researcher in the position of not having to assume any territorial limitation from what 'local' means.

The question I asked myself when developing the studies discussed in this chapter was: what it takes to study practices across concretely connected space and time? Are these the same methods adopted to study occasioned interactions—the types of brief encounters ethnomethodology has been interested in so far—or do I need a different vocabulary? The answer in this chapter has been that fundamental ethnomethodological precepts are still valid and useful.

However, my attention for people's skill in making sense of what is beyond their local reach (e.g. appresentation) also shows that to achieve a naturalistic description of members' methods in increasingly connected situations, prominent assumptions in ethnomethodological writings such as the idea of "setting" as a somewhat closed, self-compensating, self-terminating circuit of action need to be reconsidered.[5]

To put it down in Pollner's way, the 'localist turn' in the way interaction has been studied should be addressed by progressing an initiative that leads ethnomethodology to interrogate its own practices and presuppositions (Pollner 2012a). Otherwise, the risk is the uncritical reproduction of concerns that were strictly related to the historical period when "Studies in Ethnomethodology" (Garfinkel 1967) were forged.[6] These concerns do not necessarily apply equally today. On the contrary, they may prevent the empirical rendition of members' methods to achieve social order, which is at the very core of the ethnomethodological programme and its disturbing approach.

As Garfinkel incorporated accounting, accounts and accountability into the very specification of ethnomethodology's subject matter, the attempt of contemporary ethnomethodology should be to extend this effort by considering social practices based on digital platforms such as for example those occurring via social and professional networking sites. Garfinkel's homo sociologicus was an 'accountant' (Pollner 2012a: 25), making his/

[5] I recognize that ethnomethodology is not an 'unitary perspective', but rather 'a variety of distinctive subfields…[comprised] of several bodies of work, rather than a single enterprise' (Maynard and Clayman 1991: 386). Therefore, in this chapter I will focus especially on Harold Garfinkel's contribution (1967), and particularly, on the way his contribution has been reported in connection with the social study of science and technology (Rawls 2002; Lynch 1993).

[6] Some argue that certain misreadings of ethnomethodology have been perpetuated in more recent literature. In the retro-review debate on Suchman's book 'Plans and Situated Action' (P&SA) promoted by Timoty Koschman in the Journal of Learning Science (2003), it is argued for example that the problem with Suchman's critique to Artificial Intelligence (AI) stands in the relationship between plans and situated action. Her 'phraseology' (Sharrock and Button 2003: 263), her 'reference to Mead' (ibid.: 260) perpetuate, according to the critics, the idea that plan and the action are disjunct and that plans are representations which are operative only before and after action sequences.

her actions observable/reportable in a small space, involving talking rather than writing, and pointing to the world of the non-verbal signals accompanying such exchanges (Goodwin 1996).

Our homo sociologicus is a contemporary person organising his/her global social and professional affairs with the help of digital platforms leaving behind an inordinate amount of digital traces. To the same extent Garfinkel has been able (by using technologies such as tape recorded and video machines) to 'reveal levels of theoretical possibility [in the understanding of the in vivo reality of social facts: ed.] that were not apparent to Durkheim' (Rawls in Garfinkel 2002: 21), social and professional networking sites reveal unexplored levels of possibility in the detailed understanding of the integration of social facts.

Despite all this, as we will note in the next chapter (Chap. 3), it is indeed surprising how scarce the engagement of digital research scholars is with new theorizing.

As technologies that connect and synchronize multiple and dispersed scenes of action, digital social networks offer the possibility to track actors' affairs beyond what is ordinarily portrayed in natural settings. Therefore, to effectively describe the contemporary social order, ethnomethodology should address how coordination can also take place in the absence of co-located resources and how social actors work fairly hard in constituting not only a setting's feature, but also a larger infrastructure of intelligibility and possibility.

A difficult and risky task of ethnomethodology has always been to see 'work' where others see 'facts'. This applies to all interactional devices that rely on context for intelligibility, a context that is itself open-ended. The user of an expression finds and builds a usable coherence by working with the scenic and temporal circumstances of that expression. However, the novel conditions that digital technologies set for social interaction to acquire a meaning may imply that these circumstances are not given, but made to work. As one of the few social scientists who actually engage with new theorising, Karin Knorr Cetine once noted that the situation that makes the user of an expression able to respond back to the expressor in a global and digitally connected world can easily break down if not prepared, constantly updated and widely distributed: a state of affairs that she defines 'ontologically fluid' (Knorr Cetina 2009: 71). That means that in digital situations, screen realities and the algorithms that produce them become 'an interactional partner for participants' (Knorr Cetina 2009: 72). These insights will be crucial in the remainder of the book to my proposal for an ethnography of data science (see Chap. 5).

References

Abbott, A. (1992). From Causes to Events: Notes of Narrative Positivism. *Sociological Methods & Research, 20*(4), 428–455.

Callon, M. (1986). Power, Action and Belief: A New Sociology of Knowledge? In J. Law (Ed.), *Some Elements of a Sociology of Translation: Domestication of the Scallops and the Fishermen of St Brieuc Bayin* (pp. 196–223). London: Routledge.

Campagnolo, G. M., & Fele, G. (2010). From Specifications to Specific Vagueness: How Enterprise Software Mediates Engineering Relations. *Engineering Studies, 2*(3), 221–243.

Campagnolo, G. M., Pollock, N., & Williams, R. (2015). Technology as We Do Not Know It: The Extended Practice of Global Software Development. *Information and Organization, 25*, 150–159.

Ciborra, C. U. (2000). *From Control to Drift: The Dynamics of Corporate Information Infrastructures.* Oxford: Oxford University Press.

Ciborra, C. (2006). The Mind or the Heart? It Depends on the (Definition of) Situation. *Journal of Information Technology, 21*, 129–139.

Dourish, P. (2001). *Where the Action Is: The Foundations of Embodied Interaction.* Cambridge, MA: MIT Press.

Downey, G. L. (2009). What Is Engineering Studies For?: Dominant Practices and Scalable Scholarship. *Engineering Studies, 1*(1), 55–76.

Dutton, H. W. (2013). The Social Shaping of Digital Research. *International Journal of Social Research Methodology, 16*(3), 177–195.

Eisenberg, E. M. (1984). Ambiguity as Strategy in Organizational Communication. *Communication Monographs, 51*, 227–242.

Fele, G. (2009). Why is Information System Design Interested in Ethnography? Sketches of an Ongoing Story. *Ethnografia e Ricerca Qualitativa, 1*, 1–38.

Garfinkel, H. (1967). *Studies in Ethnomethodology.* Englewood Cliffs, NJ: Prentice-Hall.

Garfinkel, H. (2002). Ethnomethodology's program: Working out Durkheim's aphorism, Lanham, Rowman & Littlefield.

Goodwin, C. (1996). Transparent Vision. In E. Ochs, S. Emanuel, & S. Thompson (Eds.), *Interaction and Grammar* (pp. 370–404). Cambridge: Cambridge University Press.

Harper, R., & Carter, K. (1994). Keeping People Apart: A Research Note. *Computer Supported Cooperative Work, 2*(3), 199–207.

Henke, C.R. (1999). The mechanics of workplace order: Toward a sociology of repair. Berkeley Journal of Sociology, 44, 55–81.

Hutchins, E. (1996). *Cognition in the Wild*. Cambridge: MIT Press.

Hutchins, E., & Palen, L. (1998). *Constructing Meaning from Space, Gesture and Speech*. In Resnick, L., Säljö, R., Pontecorvo, C. & Burge, B. (eds) 1997. Discourse, Tools and Reasoning: Essays on Situated Cognition. Berlin: Springer-Verlag.

Husserl, E. (1960, 1931). Cartesian meditations. Kluwer: Cairns, D., trans. Dordrecht.

Irons, R. L. (1998). Organizational and Technical Communication: Terminological Ambiguity in Representing Work. *Management Communication Quarterly*, *12*(1), 42–71.

Kallinikos, J. (2004). Farewell to constructivism: Technology and context-embedded action. In C. Avgerou, C. Ciborra, & F. Land (Eds.), The social study of information and communication technology: Innovation, actors, and contexts. Oxford: Oxford University Press.

Knorr Cetina, K. (2009). The Synthetic Situation: Interactionism for a Global World. *Symbolic Interaction, 32*(1), 61–87.

Lave, J. (1988). *Cognition in Practice: Mind, Mathematics and Culture in Everyday Life*. New York: Cambridge University Press.

Law, J., & Urry, J. (2004). Enacting the Social. *Economy and Society, 33*(3), 390–410.

Lynch, M. (1993). Scientific practice and ordinary action: Ethnomethodology and social studies of science. Cambridge: Cambridge University Press.

Maynard, D.W., Clayman, S.E. (1991). The Diversity of Ethnomethodology. Annual Review of Sociology, Vol. 17: 385–418.

Monteiro, E., Jarulaitis, G., & Hepsø, V. (2012). The Family Resemblance of Technologically Mediated Work Practices. *Information and Organization, 22*(3), 169–187.

Monteiro, E., Pollock, N., Hanseth, O., & Williams, R. (2013). From Artefacts to Infrastructure. *Computer Supported Cooperative Work, 22*(4–6), 575–607. (CSCW).

Nicolini, D. (2011). Practice as the Site of Knowing. Insights from the Field of Telemedicine. *Organization Science, 22*, 602–620.

Ophir, A., & Shapin, S. (1991). The place of knowledge: A methodological survey. Science in Context, 4, 3–21.

Orr, J. (1996). *Talking About Machines: An Ethnography of a Modern Job*. Ithaca: Cornell University Press.

Pollner, M. (2012a). 'The End(s) of Ethnomethodology'. The American Sociologist. 43(1), 7–20.

Pollner, M. (2012b). 'Ethnomethodology from/as/to Business'. The American Sociologist. 43(1), 21–35.

Pollock, N., & Williams, R. (2009). Software & organizations: The biography of the enterprise-wide system or how SAP conquered the world. London: Routledge.

Pollock, N., Williams, R., & D'Adderio, L. (2007). Global Software and Its Provenance: Generification Work in the Production of Organizational Software Packages. *Social Studies of Science, 37*, 254–280.

Rawls, A.W. 2002. "Editor's Introduction", in H. Garfinkel, Ethnomethodology's Program. Working out Durkheim's Aphorism, Lantham, Rowman & Littlefield, pp. 1–64.

Rodemeyer, L.M. (2006). Intersubjective temporality: It's about time. Dordrecht: Springer.

Schutz, A. (1971). The Problem of Social Reality. In N. Maurice (Ed.), *Collected Papers I.* the Hague: Nijhoff.

Schutz, A., & Luckmann, T. (1989). Structures of the life-world. Vol. II, Evanston, Illinois, USA: Northwestern University Press.

Sharrock, W., Button, G. (2003) Plans and Situated Action Ten Years On. The journal of the Learning Sciences, 12(2), 259–264.

Smith, W. (2009). Theatre of use: A frame analysis of information technology demonstrations. Social Studies of Science, 39(3), 449–480.

Sommerville, I, Rodden, T., Sawyer, P., Bentley, R., & Twidale, M. (1993). *Integrating Ethnography Into the Requirements Engineering Process.* In Proc. IEEE International Symposium on Requirements Engineering (RE'93), San Diego, pp. 165–173.

Stump, David, J. 1996. 'From Epistemology and Metaphysics to Concrete Connections', in D. Stump and P. Galison (eds), Disunity of Science: Boundaries, Contexts, and Power. Stanford, CA: Stanford University Press, pp. 255–86.

Suchman, L. A. (1987). *Plans and Situated Actions: The Problem of Human-Machine Communication.* Cambridge: Cambridge Press.

Suchman, L. (1994). Working Relations of Technology Production and Use. *Computer Supported Cooperative Work, 2*(1-2), 21–39.

Suchman, L., Trigg, R., & Blomberg, J. (2002). Working Artefacts: Ethnomethods of the Prototype. *British Journal of Sociology, 53*(2), 163–179.

Tingling, P., & Parent, M. (2004). An exploration of enterprise technology selection and evaluation. Journal of Strategic Information Systems, 13, 329–354.

Vaast, E., & Walsham, G. (2009). Trans-Situated Learning: Supporting a Network of Practice with an Information Infrastructure. *Information Systems Research, 20*(4), 547–564.

Wittgenstein, L. (1953). Philosophical investigations. Oxford, UK: Blackwell.

Wittgenstein, L. (1978). Remarks on the foundations of mathematics (3rd ed.). Oxford: Basil Blackwell.

CHAPTER 3

The Analogue Mapping

Abstract This chapter addresses the role of visualisation in social data science by offering a comparison between situational maps and data-driven topological network approaches. Through a variety of case studies, it discusses the methodological politics involved when resource-intensive computational solutions are part of the research design. It suggests rethinking the foundations of data science projects by shifting terms from certainty to ambiguity and from value-neutral to intrinsically socio-political and it extends this critique to the graphical means of expressing knowledge in data science. It describes the overall digital social science programme as lacking ambition and failing to challenge existing sociological concepts.

Keywords Situational analysis • Linked ecologies • Data visualisation • Social network analysis

In Chap. 2, the distinctive issue was the recognition that objects and settings have a networked character, which poses particular conceptual challenges for phenomenological sociology notions. This chapter addresses the expansion of the object of study methodologically. It will explore social research methods that are predicted on the basis of providing a more strategic approach to the extendedness of technology-mediated practices and of the data that we have about them.

G. M. Campagnolo, *Social Data Science Xennials*,
https://doi.org/10.1007/978-3-030-60358-8_3

35

These approaches will be exemplified by their application to two cases: one is the case of the early introduction of packaged enterprise software in the Italian public sector over a ten-year period and across a number of interconnected field sites. The other is the development over five years of a digital infrastructure for intellectual property trading in the UK. While the previous chapter dealt with the temporal limitations of ethnographic approaches, this chapter addresses its space constraints. It also begins to address the role of visualisation in social data science by offering a comparison between situational maps and data-driven topological network approaches. In conclusion, it will develop a critique of the contemporary use of social network approaches in digital social research. The overall programme of digital approaches to social science will be described as that of showing more clearly the validity of old theories. It will be argued that this agenda lacks ambition and fails to challenge existing sociological concepts.

As part of a continued effort to conceive personal intellectual biographies as an integral part of method, in what follows I will begin by accounting for my experience of attending larger-scale technological projects and how this experience marked my approach to social data science. Both the case studies presented in this chapter cover technological projects that vastly exceeded the planned duration and included apparently inexplicable misjudgements regarding the most appropriate technological solution. Similar to what was observed in Chap. 2 regarding vagueness (Campagnolo and Fele 2010) and ambiguity (Irons 1998), this chapter helps showing that though technological development is often assessed from a narrow instrumental perspective which focuses exclusively on whether the planned technical outcomes are achieved, the understanding of many technological projects on the ground requires a different perspective—one that explores how tolerance to failure might result when an initiative becomes subject to radically different tests of adequacy by the players involved. I will show how this experience helped me to understand in more concrete terms notions such as slow science (Stengers 2011) and non-positive quantitative approaches (Bouillier 2018), informing my approach to the application of data science to social science.

SOME CASE STUDIES OF ANALOGUE MAPPING

The first case derives from my experience as a consultant for the Italian Government Digital Agenda. The Centre for Informatics in the Public Administration (CNIPA) initiated an ambitious project to improve interoperability across sectors (e.g. health care and social care) and across local authorities. At the time, the two main architectural models to do this were the institutional interface architecture (INA) and the service provider architecture (SPA), with the former consisting in a distributed set of plug-ins and the latter relying on one single service provider. Despite empirical evidence collected over a period of 20 years demonstrating the viability of the latter (Hanseth and Bygstad 2012), CNIPA decided to adopt the institutional interface architecture (INA) approach.

The question I investigated in this case was: why if SPA-based architectures seem to be more successful, larger amount of government funding was directed towards INA type of architecture development instead?

The story goes that at national level, a programme has been established that essentially mandates that peripheral organisations (regional authorities) should adopt INA-based architectures. The initial conjecture was that this architectural model allowed actors at higher level to play a more active role by establishing nation-wide interoperability standards. However, the move towards INA architectures was also welcomed by the 20 Regional Authorities participating in the programme. The reason for it is that Regional Authorities' in-house IT companies acting since then as service providers in SPA type of architectures were perceived as underperforming. Suffering from the 'red tape' syndrome, which is common to large bureaucratic organisations, they were underservicing local municipalities. Therefore, regional authorities recognised in INA types of architecture the opportunity at once (1) to get rid of the blockages generated by the in-house IT company acting as service provider in SPA type of architecture and (2) to foster the regional 'system of innovation' made of SME IT companies most of them spinning off from a vibrant IT research scene (IT University and a number of IT-oriented research foundations are common in most regions). Despite the premises to use the architectural control point of national standards to ensure a fairer competition and growth of the local IT market, what happened is that the co-operation based light-weighted INA type of architecture did not stand the test of time. The key problem was that the organisational architecture had a 'hollow core'. The idea to work with a virtual central actor (i.e. not backed up by an existing organisation) and put it in charge to develop and coordinate the infrastructural standards and then delegate the installation of the necessary plug-ins for data exchange to peripheral organisations (municipalities and health care departments) proved too fragile.

My conclusion was that large nationally (or internationally) funded projects based on INA type of architectures, although underperforming according to many existing technological criteria (bootstrapping, organisational mirroring, etc…) offered control points for regulatory organisations like Regional Authorities to control the IT market. Although demonstrably less bootstrappable, adoptable, adaptable or extensible, and weaker in terms of the 'organisational mirroring' hypothesis, INA architectures offered higher-level actors stronger control points.

What emerged was that in projects such as this, it is not computational possibilities alone which determine outcomes. Computational or infrastructural methods and approaches must be conceived in their context and in reciprocal terms. This experience and the need to express this finding in more analytical terms prompted the use in subsequent studies of situational analysis (Clarke 2005) and the linked ecologies approach (Abbott 2005).

SITUATIONAL ANALYSIS

In this section I present how I applied situational analysis (Clarke 2005) in my study of another technological project in the public sector (Campagnolo 2013). This was a ten year-long study of Enterprise Software implementation in a large public sector organisation. In this case, the contextual factor was the role of IT consultants. Four different types of consultants have been hired to manage the implementation of an enterprise system and the responsibility for the relationship with consultants migrated over time to various different organisational actors. Furthermore, as with the evolution of the structure of IT expertise, particular historical contingencies—such as new regulations about access to consultancy—affected the configuration of client–consultant relationships. The case raised questions as to whether consultants organise technological fields or consultancy is shaped by key negotiations and the situation. To answer this question, I focused on the way key mediations between different types of management consultants specialising in IT and their local government client changed several times during the period under scrutiny (see Fig. 3.1).

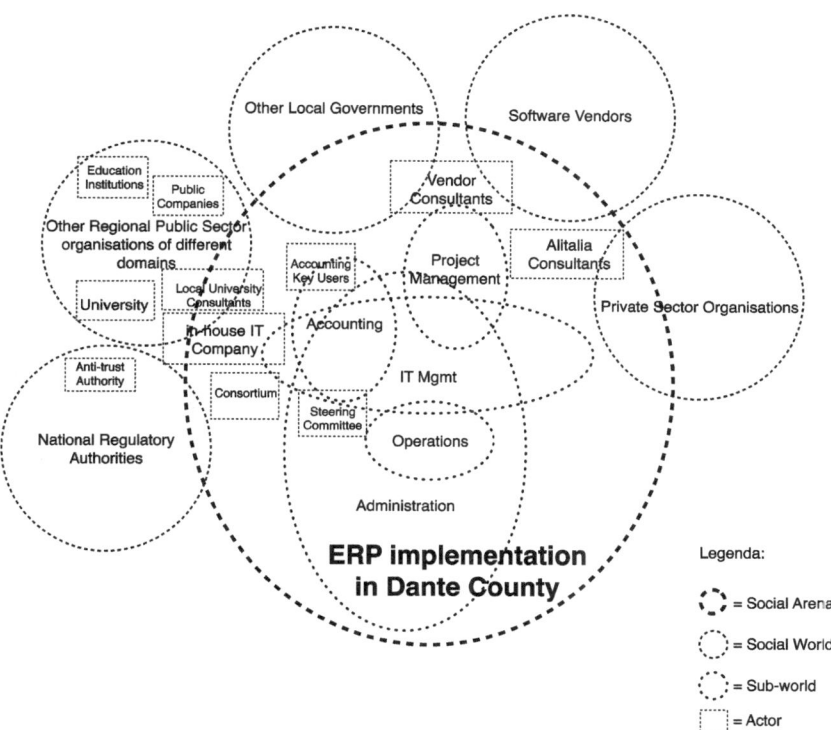

Fig. 3.1 This is a situational map I developed as part of a study on the implementation of enterprise software in the public sector. It is meant to show how the situation under scrutiny is populated by a variety of other organisations that, to different extents, contribute to shaping ERP implementation in a particular setting/organisation. There are other local organisations from different public service domains and other local government organisations from different regions that are engaged in similar technological projects and share participation in the social world/arena

The use of situational maps (Clarke 2005) for the analysis of this case provided insights into the deeply contextual reflexivity between client–consultant relationships and the evolving situation in the public sector domain. During the same technological project, clients changed consultants multiple times according to previous implementation experiences, intra- and inter-organisational dynamics and the evolving situation of the

IT market in the public sector. The case helps elucidate how the formation of client–consultant relationships is partly dependent upon internal organisational dynamics: decisions are highly dependent upon changes in relevant organisational social groups and their respective relationships with those who managed consultants' relations prior to them. However, the case also directs attention to the relevance of inter-organisational dynamics and to the pattern of diffusion of the enterprise software across different sectors as well as the relevant inter-organisational relations between authorities across local, regional and national boundaries.

The situational analysis approach used in this case is a grounded theory approach that offers three main cartographic approaches:

1. Situational maps that lay out the major human, non-human, discursive and other elements in the research situation of inquiry and provoke analysis of relations among them
2. Social worlds/arenas maps that lay out the collective actors, key non-human elements, and the arena(s) of commitment and discourse within which they are engaged in ongoing negotiations—meso-level interpretations of the situation
3. Positional maps that lay out the major positions taken, and not taken, in the data vis-à-vis particular axes of difference, concern, and controversy around issues in the situation of inquiry.

All three kinds of maps are intended as analytic exercises, fresh ways into social science data that are especially well suited to contemporary studies from solely interview-based to multi-sited research. They are intended as supplemental approaches to traditional grounded theory analyses that centre on action—basic social processes. Instead, these maps centre on the situation of inquiry. Through mapping the data, the analyst constructs the situation of inquiry empirically. The situation per se becomes the ultimate unit of analysis and understanding its elements and their relations are the primary goals.

Situational analysis offers ways to represent connections among apparently unrelated organisational settings across different generations of discourse production. The grounded approach of situational analysis is based on the development and continuous refinement of situational maps: maps that specify what there is empirically in considerable detail and from multiple angles of perception.

Situational analysis also conveys an understanding of the intra-organisational and inter-organisational space as meaningful for the case in analysis. While current interactional analyses occur across cases that are assumed not to have any relation with each other, situational analysis allows the researcher to progressively find and build connections between apparently disparate cases.

Linked Ecologies

Another example where the context of the technological project encompassed an even wider array of actors is a study conducted with colleagues Hung The Nguyen and Robin Williams on the development over five years of the Copyright Hub: a digital infrastructure for Intellectual Property trading being developed in the UK (Campagnolo et al. 2019). Similarly to the above case, one key puzzle posed by this study is why would the Copyright Hub opt for a technological solution (i.e. a data model) that is seen as not ideal (i.e. too complex and 'verbose') for creating a new infrastructure for which uptake is key? As per the above case on service architectures, if we only address the outcomes as evaluated by an external audience at the end of the development process we cannot make sense of this.

There is one more specific reason why the lesson learnt from this latter case translates more directly into developing our understanding of social data science. It shows the methodological politics involved when resource-intensive computational solutions are part of the research design. The Copyright Hub project is indeed an example of a government–industry partnership for digital innovation where translational research intermediaries (the Digital Catapult in this case) were putting commercial initiatives in touch with relevant, albeit unproven, research findings on data science (i.e. data modelling for data interoperability).

In order to analyse this, we refer to Andrew Abbott's linked ecologies framework (Abbott 2005). The ecological argument is used in sociology to acknowledge an intermediate account between individualist and emergentist approaches to social processes. It recognises the agency of individual actors and also the influence of larger structures and relationships which pattern the choices available to individuals. In a linked ecologies approach, actors aren't only dependent upon the interaction with other actors in the same ecology. The outcome of actors' interaction within 'one' ecology also triggers and becomes subject to events occurring in neighbouring ecologies. In a linked ecologies understanding, the outcome of, say, a technological project is not defined by a set of external, univocal or fixed criteria. To succeed in 'one' ecology a technological development must provide results also to allies in adjacent ecologies, despite 'mismatches of the rhythms of interests between ecologies' (Abbott 2005, 264). Strategies that provide these kinds of dual (or multiple) rewards, Abbott calls 'hinges'. A hinge is a strategy that works in both ecologies at once but it can be perceived in fundamentally different ways in the two (or more) ecologies that it links. A linked ecologies approach therefore provides tools for understanding contextual influences over the dynamics of development of technological innovations and their related promises. Describing technological innovations as 'hinges' helps realise that the evolution of the Copyright Hub as a market actor should be understood also in relation with neighbouring professional ecologies and the political ecology of the UK government as well as a project which fosters the interest of academic actors such as research institutions.

The key question thus becomes: what is the payoff that the application of the Linked Data Model to the Copyright Hub provides not only to SMEs in creative industries, but also to established market actors, the UK government and the more or less academic 'movements' advancing digital rights management ideas?

As the most recent re-incarnation of a long-standing interest shared by academics and activists alike to identify viable solutions for digital right management, the application of the Linked Data Model to the Copyright Hub can be considered a successful 'hinge' action—despite its 'verbose' data model—in that it gave creative industries a way to coalesce that would not simply be discounted as a lobbying. Reciprocally, the Copyright Hub gave Linked Data Model advocates a new home to pursue their 'academic' programme of building interoperability standards and metadata for copyright exchange. While the clear advantage from the policy makers—who once described the IP framework 'as a priesthood [...] enacted by these quite funny men of a certain age in legal chambers, dusty files all around them'—was to delegate the solution of the IP framework controversy to technological innovation, as opposed to more legislation.

By referring to the INA architecture and the Copyright Hub data model examples I suggest to rethink the foundation of resource-intensive computational solutions such as those used in many contemporary data science projects by shifting terms from certainty to ambiguity and from value-neutral to intrinsically socio-political. The lesson learnt from the idea of situationally reciprocal gains, which emerges from the linked ecologies and the situational analysis approaches, should be extended to any form of data production in data science as fundamentally co-dependent, relationally constituted between, among others, the observer and the observed phenomena.

In the next section I extend my critique to the graphical means of expressing knowledge produced by data science as an interpretative complex undertaking.

COMPARING SITUATIONAL AND DIGITAL NETWORK MAPS

So far in this chapter I have discussed the application of analogue mapping methods such as situational analysis (Clarke 2005) and the linked ecologies approach (Abbott 2005) whose key point is to offer ways to represent connections among apparently unrelated organisational settings or even sectors and support the understanding of changes in discourse and actors over time. These approaches have been exemplified by their application to two cases. One is the case of the early introduction of packaged enterprise software in the Italian public sector over a ten-year period and across a number of interconnected field sites. The other followed the development over five years of a digital infrastructure for intellectual property trading in the UK.

I now proceed to the second part of the chapter, which is dedicated to the ways of representing extended knowledge by comparatively looking at situational maps and digital network maps.

Comparing different types of maps pre- and post-digital maps, including issue maps, has already been attempted by Munk and Jensen (2014). While the difference between different forms of networks—actor-networks, social networks and digital network—has been discussed in Venturini et al. (2019). In consideration of what has been already discussed in these studies, in this section I will try something new and different. Given that digital network maps gained currency as one of the main tools of digital social research (Rogers 2013; Rogers and Marres 2000), I want to extend the scope of the comparative exercise to encompass other social research mapping methods such as situational maps.

I will show how analogue situational maps allow and cater for a much wider variety of actants. While network nodes tend to be of the same type, situational maps as those presented in this chapter allow representing entities as different as social actors, social worlds and social arenas. I will also discuss how the researcher is the only instrument that mediates situational maps; while digital maps rely heavily on digital infrastructures. Using my studies of expert work in the IT sector as an example, I will list a number of areas where the mediation of digital infrastructures is apparent.

I will begin by discussing the limitations linked to the fact that digital infrastructures can only capture interactions that are mediated by digital (social) media.

Factors that affect the typology of interactions that can be mediated by social media data include the way cognitive authority is established in a field. Not all types of experts have the same legitimators. At first glance, as sellers of advice, the IT engineers and consultants I have described so far as part of my studies might appear to belong to the type of experts whose personal expertise is tested by individuals (i.e. the buyers of their expertise). In such a case, more (Twitter) followers might legitimately mean more cognitive authority. However, I learned from fieldwork that IT expertise is legitimated by particular types of buyers: government or corporate bureaucrats with discretionary power (Chief Information Officer—CIO, Chief Marketing Officers—CMO), who are experts in their own right and whose views are accepted as authoritative by Executives and their Boards. Turner defines as sectarian (Turner 2001: 138) those experts whose validating audience does not correspond with the general public. As such, their influence is hardly measurable from simply surveying any overt measure of association (e.g. a vote, a like, number of followers, a re-tweet).

Exhibiting a non-representative sample (Duggan 2015), that is, a sample that does not reflect how cognitive authority is established in a field, is not the only way digital infrastructures mediate research results. Social science scholars also repeatedly challenge the 'naturalness' of the data generated by digital platforms, recognising that digital platforms such as social media are non-neutral (Coleman 2012; Gillespie 2010). Internet-generated data such as those mostly used to generate digital network maps might give near real-time access to new social groupings (Housley et al. 2014). However, platform data are low-fidelity in that digital data stand out for its artificiality (Marres 2017: 52; Halavais 2013). As Shaw (2015) points out, social media data are strongly marked by platform effects, such as search terms suggested by auto-complete functions or users taking up hashtags that are trending on the platform in question.

So far, I have discussed aspects related to the apparatus that generates the data used as input for digital network maps. My observations regarding digital maps have been comparative (vis-à-vis to analogue maps) as opposed to necessarily critical.

A further point concerns digital network maps as a way of representing extended knowledge. As Venturini et al. (2019) point out, with digital network maps one can only represent what is mathematically tractable. Conventional graph mathematics allows some nodes differentiation (Guillaume and Latapy 2006) while when it comes to edges the homogeneity is highly imposing. The limitation is sometimes referred to as the problem of 'parallel edges': the difficulty to compute networks in which nodes can be connected by more than one edge at a time as in when one tries to project different types of relations between two nodes—or traces coming from different media—on the same network (cfr. Rieder 2013).

Finally, and more problematically, there is an irreversible essentialism in the way network approaches display knowledge. To reiterate Venturini et al. (2019): with their resemblance to representations used to manage the flow of transportation, they contribute to reproduce the realist idea that phenomena are observer-independent, that is, that society has in fact become a network (cfr. Castells 2000; Van Dijk, J. 2012). Through the network graph the researcher assigns fixed properties to the nodes and the edges that are mapped in the network. Furthermore, these epistemological preferences remain concealed under a guise of completeness. Networks become automatically complete with what is available in the data. There is no way by looking at them to have a sense of areas where lack of data makes the network incomplete.

Transparency, completeness and independence are not the assumption with situational maps. They aptly manifest the traditional reluctance in social research to pin down methods too tightly or explicitly. Their low fidelity is a signal that situational maps cannot ossify into an instrumentalist account. They only help understand collective topologies, not replace them. It helps reveal the constructed-ness of the description, in line with the basic sociological assumption of the critical distance between the observed world and its interpretation.

The comparative examination of digital network visualisation and low-key analog maps undertaken in this chapter is a reminder that no matter the amount of data or the precision of the computational engine, in social data science it is important to recognise that knowledge is constructed, taken, not simply given as a natural representation of pre-existing facts.

In Chap. 4 I will observe possible solutions towards the critical points raised in this chapter with respect to digital network maps, their apparatus for producing data and display knowledge. The proposed solution will explore ways of using analogue and digital methods in tandem in the context of the same research project. Before that, I want to conclude the chapter with a reflection regarding the sociological theories that are often found in conjunction with a research design based on digital methods and tools.

New Methods: Old Theories

The idea in this chapter has been that of comparing analogue and digital ways of representing complex, interactive narratives involving several independent actors by coding and visualising extended ethnographic narratives.

When it comes to theory development in most of the studies that readily adopted digital methods (including digital maps), it is often lamented that classical questions of social order are the result of a sociological imagination that was limited by lack of data (i.e. digital data) or methods (i.e. digital methods). For example, Latour et al. (2012) say that the classical distinction between individual and aggregate levels is the legacy of a time where the possibility of navigating digital data sets of network relations wasn't as complete and as quickly accessible as it is today.

However, the re-orientation of sociological imagination towards a flat ontology is not necessarily the only possible outcome of the encounter with the digital. The point of this book and of the comparison made in this chapter between analogue and digital maps is that such understanding predates digital traces.

More disappointment comes for those who expect that the coming together of social science and data science should be the occasion to generate new theories that are neither pre-existing computational approaches applied to social science, nor pre-existing social theories made testable by the advent of new data. Quite ironically, what authors such as Latour et al. (2012) make of the newly available techniques of digital navigation and visualisation is the opportunity to give a new career to old theories. Latour et al. (2012), for example, identify Tarde's idea of monad (Tarde 1895) as the best candidate to digitally test alternative definitions of atoms, interactions and structure. Perhaps the lack of originality is due to the authors neglect of what 'quantitative' analysis can offer beyond the visual network analysis deployed in their account. As Moats and Borra (2018) aptly recognise, Latour et al. (2012) do not fully address variables or causality, nor do they even use the numerical properties of networks, as computational social network analysts would, to make statistical claims about the centrality of particular nodes. The networks are primarily interpreted visually to spot patterns and identify interesting entities and relationships (Venturini et al. 2019), quantifications are generally only used to cluster, filter and spatialise networks. So 'quantitative' here has more to do with computer-assisted techniques than with quantifications and measurement per se.

It must be said, however, that Latour et al. (2012) are not alone in using new methods to test old theories. In his paper on the contemporary relevance of the Chicago School, Abbott (1997) also said that by looking at these new analytic techniques one can realise that the old Chicago died because its theory was too far ahead of its methods (Abbott 1997: 1177). Both Latour and Abbott have something in common with a more descriptive tradition in quantitative methods and their approach can be seen in relation to the shortcomings of the 'variable' and 'causal' paradigms in mainstream sociology. More surprising is to find reliance on old sociological theories by the flag-bearers of the transformative role that data science can have for social and anthropological sciences (Seaver 2014; Castelle 2018). In the call for paper of a recent workshop held at the Copenhagen Centre for Social Data Science, Pedersen et al. (2020) mention that the idea of what they call machine anthropology reflects the Lévi-Strauss structural idea (Lévi-Strauss 1963). While in his work on the social lives of Generative Adversarial Networks (GAN), Castelle explores the isomorphism between Bourdieu's concept of habitus (Bourdieu 1977) and the architecture of one of the most advanced machine learning techniques. As necessarily implied by the comparison, the argument is that deep learning research is re-discovering and re-inventing poor variations of old social theories.

Perhaps the attitude towards old theories is a defensive one: to guard the discipline against people with little or no training in anthropology taking on big socio-cultural questions with computational methods. As Nick Seaver says, 'we find PhDs in Electrical Engineering trying to algorithmically define musical genres, computer scientists modeling family ties in social networks, and autodidact software developers designing "content discovery" apps around their own theories of cultural influence and flow'

(Seaver 2014). The reason for reviving social theories of the time past is perhaps also that authors want to distance themselves from 'epochalist' thinking (Savage 2009) which sees social data science as bringing significant theoretical change in the discipline.

However, it is very apparent that neither Tarde, nor Lévi-Strauss or Bourdieu have ever worked in methodologically interdisciplinary projects including computational methods.[1] These cannot be examples of the empractised and idiosyncratic approach to social data science we are advocating.

As suggested by Moats and Borra (2018) and by Collins (1994) before them, the instinctual wariness of the so-called qualitative researcher towards digital research is more a matter of appropriation of genealogies of research technologies than any failure to adhere to a scientific epistemology. A debate where new research technologies are used to dispute the interpretation of older discoveries is only helping to reproduce the fundamental disability of social sciences to appreciate the need to produce new phenomena and adhere to the rapidly moving research front associated with the appropriation of new research hardware. The reasons for the instinctual wariness of the qualitative researcher towards digital research should be found in a current situation where ready-made tools are parachuted in from nowhere, what Moats and Borra (2018) describe as distant tool design. This can only be resolved by a better understanding of how automated tools and qualitative analyses can work together concretely and idiosyncratically around particular empirical cases with specific data formats. This is what I am going to do in Chap. 4.

REFERENCES

Abbott, A. (1997). Of Time and Space: The Contemporary Relevance of the Chicago School. *Social Forces, 75*(4), 1149–1182.

Abbott, A. (2005). Linked Ecologies: States and Universities as Environments for Professions. *Sociological Theory, 23*(3), 245–274.

Bouillier, D. (2018). Médialab Stories: How to Align Actor Network Theory and Digital Methods. *Big Data & Society, 5*(2), 1–13.

Bourdieu, P. (1977). *Outline of a Theory of Practice.* Cambridge: Cambridge University Press.

---

[1] Although Bourdieu's team fieldwork used a mix of qualitative/ethnographic and statistical approaches, in particular Multiple Correspondence Analysis (MCA) and other Geometric Data Analysis (GDA) techniques.

Campagnolo, G. M. (2013). The Evolution of Client-Consultant Relationships: A Situational Analysis of IT Consultancy in the Public Sector. *Financial Accountability & Management, 29*(2), 0267–4424.

Campagnolo, G. M., & Fele, G. (2010). From Specifications to Specific Vagueness: How Enterprise Software Mediates Engineering Relations. *Engineering Studies, 2*(3), 221–243.

Campagnolo, G. M., Nguyen, H., & Williams, R. (2019). The Temporal Dynamics of Technology Promises in Government and Industry Partnerships for Digital Innovation: The Case of the Copyright Hub. *Technology Analysis & Strategic Management, 31*(8), 972–985.

Castelle, M. (2018). *The Social Lves of Generative Adversarial Networks*. Conference on Fairness, Accountability and Transparency (FAT* '20) January 27–30, 2020, Barcelona, Spain. ACM, New York, NY.

Castells, M. (2000). *The Rise of the Network Society*. Oxford: Blackwell.

Clarke, A. E. (2005). *Situational Analysis: Grounded Theory After the Postmodern Turn*. Thousand Oaks, CA: SAGE.

Coleman, G. (2012). *Coding Freedom: The Ethics and Aesthetics of Hacking*. Princeton: Princeton University Press.

Collins, R. (1994). Why the Social Sciences Won't Become High-Consensus, Rapid-Discovery Science. *Sociological Forum, 9*(2), 155–177.

Duggan, M. (2015). *Mobile Messaging and Social Media 2015*. Pew Research Center. Retrieved from http://www.pewinternet.org/files/2015/08/Social-Media-Update-2015-FINAL2.pdf.

Gillespie, T. (2010). The Politics of Platforms. *New Media and Society, 12*(3), 347–364.

Guillaume, J., & Latapy, M. (2006). Bipartite Graphs as Models of Complex Networks. *Physica A: Statistical and Theoretical Physics, 371*(2), 795–813.

Halavais, A. (2013). Structure of Twitter: Social and Technical. In K. Weller, A. Bruns, J. Burgess, M. Mahart, & C. Puschmann (Eds.), *Twitter and Society* (pp. 29–42). New York: Peter Lang.

Hanseth, O., & Bygstad, B. (2012). *ICT Architecture and Project Risk in Inter-Organizational Settings*. ECIS 2012 Proceedings. 130.

Housley, W., Procter, R., Edwards, A., Burnap, P., Williams, M., Sloan, L., & Greenhill, A. (2014). Big and Broad Data and the Sociological Imagination: A Collaborative Response. *Big Data and Society, 1*(2), 1–15.

Irons, R. L. (1998). Organizational and Technical Communication: Terminological Ambiguity in Representing Work. *Management Communication Quarterly, 12*(1), 42–71.

Latour, B., Jensen, P., Venturini, T., Grauwin, S., & Bouillier, D. (2012). 'The Whole Is Always Smaller Than Its Parts'—A Digital Test of Gabriel Tardes' Monads. *The British Journal of Sociology, 63*(4), 590–615.

Lévi-Strauss, C. (1963). *Structural Anthropology*. New York: Doubleday Anchor Books.

Marres, N. (2017). *Digital Sociology*. Cambridge: Polity Press.

Moats, D., & Borra, E. (2018). Quali-Quantitative Methods Beyond Networks: Studying Information Diffusion on Twitter with the Modulation Sequencer. *Data & Society, 5*(1), 1–17.

Munk, A., & Jensen, T. E. (2014). Revisiting the Histories of Mapping: Is there a Future for a Cartographic Ethnology? *Ethnologia Europea, 44*(2), 31–47.

Pedersen, M. A., Albris, K., Munch Gregersen, E., & Otto, E. I. (2020). *Machine Anthropology. A Two Day Interdisciplinary Workshop*. Copenhagen: Centre for Social Data Science (SODAS).

Rieder, B. (2013). *Studying Facebook via Data Extraction: The Netvizz Application*. Proceedings of WebSci '13, the 5th Annual ACM Web Science Conference, pp. 346–355.

Rogers, R. (2013). Internet Research: The Question of Method. In C. Hine (Ed.), *Virtual Research Methods* (pp. 213–238). London: Sage.

Rogers, R., & Marres, N. (2000). Landscaping Climate Change: A Mapping Technique for Understanding Science and Technology Debates on the World Wide Web. *Public Understanding of Science, 9*(2), 141–163.

Savage, M. (2009). Against Epochalism: An Analysis of Conceptions of Change in British Sociology. *Cultural Sociology, 3*(2), 217–238.

Seaver, N. (2014). *Computers and Sociocultural Anthropology*. Savage Minds. Retrieved from https://savageminds.org/2014/05/19/computers-and-sociocultural-anthropology.

Shaw, R. (2015). Big Data and Reality. *Big Data and Society, 2*(2), 1–4.

Stengers, I. (2011, December 13). *"Another Science is Possible" A Plea for Slow Science*. Faculté de Philisophie et Lettres, ULB. Inaugural Lecture Chair Williy Calewaert 2011–2012 (VUB).

Tarde, G. (1895). *La logique sociale*. Paris: Félix Alcan.

Turner, S. (2001). What is the Problem with Experts? *Social Studies of Science, 31*(1), 123–149.

Van Dijck, J. (2012). *The Network Society*. London: Sage.

Venturini, T., Munk, A., & Jacomy, M. (2019). Actor-Network VS Network Analysis VS Digital Networks Are We Talking About the Same Networks? In D. Ribes & J. Vertesi (Eds.), *DigitalSTS: A Handbook and Fieldguide* (pp. 510–523). Princeton, NJ: Princeton University Press.

CHAPTER 4

Participative Epistemology

Abstract This chapter addresses the precise procedures and operations which are embedded in the situated encounter of ethnographic, computational and statistical methods, and asks how text mining and sequence analysis can be used ethnographically, with an attention on the mutual constitution of the fieldwork and the field. Symmetrically, it also interrogates how data science brings its own modifications to the practices of empirical sociology. The chapter argues that it is not possible to assess the prowess of a particular method in isolation or compare it with another when used in different projects/contexts. It contributes to the quali-quantitative methods debate by empirically exploring the tension between methods when they co-occur in the same study or research programme.

Keywords Participative epistemology • Sequence analysis • Text mining • Career study

In Chaps. 2 and 3 I considered the theoretical, epistemological and onto-logical sensibilities involved in a genealogical account of how a social scientist can gradually get to appreciate digital research methods. There I addressed the wide variety of existing social research approaches—not least phenomenological concepts themselves—that can help account for the relational and spatio-temporally distributed nature of social data. This chapter continues along the lines of a 'confidential', first-person narrative

G. M. Campagnolo, *Social Data Science Xennials*, https://doi.org/10.1007/978-3-030-60358-8_4

about research methods to describe my own experience of putting social data science into practice. After having made the effort in previous chapters to place my own trajectory in the long-term disciplinary relationships with phenomenological sociology and expertise in computer science associated with digitalisation, I now add my own experience with the narrative positivism (Abbott 1992) linked with the use of statistics in social research.

A criticism I made in the previous chapter of science and technology studies' ventures into data science is that they were just addressing the descriptive potential of quantitative approaches: their aesthetic register (Savage 2015). In this chapter, I argue that the debate should focus more closely on the precise procedures and operations embedded in methods themselves, rather than a more general meta-critique. What is evidently true for overtly descriptive methods such as social network analysis can be also true for apparently linear methods (Abbott 1992): quantitative claims such as variables or causality (and their quick but partial ways of summarising processual matters) can also be recast in a non-positivist frame. In this chapter the question will be: how can text mining and sequence analysis be used ethnographically, that is, with an attention on the researcher himself/herself as an instrument, with their own agenda/stakes/position/interests; in an adaptive, situated, indexical, ad hoc, just-so, iterative, exploratory, non-positive way, with an attention on the mutual constitution of the fieldwork and the field?

The other main criticism in Chap. 3 addressed the adoption of old theories as a way for sociology to retain a privileged position over the interpretation of social phenomena. To unlock what I see as a disciplinary politic of jurisdiction, I argue here that the negotiation of who may say what about society is not a new problem.[1] What's new is that the 'found' character of most big social data casts these negotiations within a larger constellation of disciplinary and extra-disciplinary collaborations. The idea of front-loading sociological concepts into data science projects contributes to solidify the disciplinary role that each intellectual party plays in the programme and reduces the possibilities for folding insights across epistemological domains. One of the not-as-yet-played-out narratives in social data science is the agonistic-antagonistic mode of interdisciplinarity (Barry and Born 2014: 12). This kind of interdisciplinarity stems from a commitment to transcend

---

[1] Boyer (2014) identifies Radin's work (Radin 1927) as the initiator of the so-called reflective turn.

the given epistemological assumptions of specific historical disciplines, a move that makes the new interdiscipline irreducible to its antecedents. As well as antagonising instrumental expectations by the industry and the narrowly mechanical understanding of the social by collaborating engineering disciplines, this means constituting the new field as antagonistic to existing sociological approaches to the study of the same object (Randall et al. 2005). It is fair to say that of these antagonisms, the least practised is the one that reflectively addresses ethnography's own presuppositions. As a further contribution to this line of enquiry, my analysis suggests that following how social data science is empractised means interrogating how it brings its own modifications to the processes and practices of empirical sociology.

Conceptual work raised some important issues on the ways qualitative and quantitative modes of inquiry complement one another in the new fields of digital social research and computational social science. However, the mixed methods literature so far has not picked up the challenges and opportunities of large-scale digital data, and is therefore at risk of remaining trapped within an obsolete distinction between qualitative and quantitative (i.e. survey-based) approaches (Blok and Pedersen 2014: 3).

Candidate concepts to frame the epistemic trouble (Marres and Weltevrede 2015) generated by the encounter of social and data science are consilience (Wilson 1998), abduction (as in Kitchin 2014) and symphonic social science (Halford and Savage 2017).

Consilience—that is, convergence of evidence from multiple, independent, and unrelated sources, leading to strong conclusions—is gradually gaining currency in the field of social data science after being proposed by George et al. (2014) in their editorial article of *Academy of Management Journal on Big Data and Management*. Quoting from Wilson they say: 'The Consilience of Inductions takes place when an Induction, obtained from one class of facts, coincides with an Induction, obtained from another different class. This Consilience is a test of the truth of the Theory in which it occurs.' (Wilson 1998).

This call for multiple forms of knowledge construction in data science is complemented by Kitchin's claim for abductive reasoning in big data analytics (Kitchin 2014). Beyond inductive or deductive approaches, abductive reasoning (Bateson 2002 [1979]) focuses on the unfolding interplay between data, method and theory. In this approach, if theory and methods are initially framing analysis, data are used to feed back into the initial framework, making the interpretation process iterative.

The symphonic social science approach put forward by Halford and Savage (Halford and Savage 2017) summarises these ideas in a claim for overcoming scepticism about big data analytics, thus liberating sociology from its defensive position. The approach suggests an end-to-end model of disciplinary integration, with a particular focus on data visualisation as a mean for data to resurface at different points in the research process (Healy and Moody 2014).

My argument, however, is that it is not possible to assess the prowess of a particular method in isolation or compare it with another when used in different projects/contexts as suggested by Wilson's concept of consilience. The contribution I want to make is to empirically explore the tension between analogue and digital methods when they co-occur in the same study.

The fact is that mixed methods research combining ethnographic and data science approaches is still rare (Blok and Pedersen 2014: 3). Detailed, first-person narratives about research methods in this particular form of contemporary interdisciplinarity are hard to find (Sandvig and Hargittai 2015: 2). In this chapter, I discuss lessons learnt from participant observation to one of these rare projects where ethnography, text mining and sequence analysis techniques have been used in combination.[2]

A Digital Sociology of Professions

The project is a social data science study of a particular form of IT expertise: that of industry analysts. Contributing to develop the first extended academic study of IT industry analysts (Pollock and Williams 2016), I came to this project after having completed over 30 interviews, undertaken over 100 hours of observation at conferences, listened and participated in more than 20 webinars and after having engaged in dozens of informal discussions with IT industry analysts. The aim of my involvement

---

[2] The issue of when we can say that methods actually co-occur is a tricky one. Take for example the Copenhagen Social Networks Study described by Blok and Pedersen (Blok and Pedersen 2014). There an interdisciplinary group of more than 25 scientists attempted to understand social interaction among the freshmen class ($N > 1000$) of the Danish Technical University. In the project, the digital mapping of social media interactions and the ethnographic fieldwork took place in two separate work-packages. Tellingly, the notion of complementarity the two authors take from Bohr suggests 'the information regarding the behaviour of one and the same object' to be taken 'under mutually exclusive experimental settings' (Bohr 1957 [1938], translated by Blok and Pedersen 2014).

was to see how the collective study of career data crawled from the web could complement the ethnographic insights that I gained from the large spectrum of ethnographic studies on IT expertise (system engineers, sales managers, management consultants, innovators) discussed in Chaps. 2 and 3. In particular, I wanted to see how the collective study of career data crawled from the web could generate additional insights to one of the fundamental questions deriving from the antecedent ethnographic study: how do people become an IT expert? Given the absence of institutionalised professionalisation systems—that is, regulated career progression—could large volumes of internet-generated data concerning work experience help understand how IT analysts gain the credibility that makes their expertise so highly reputed? The ethnographic study I contributed to carry out pointed out that IT industry analysts in their career go through 'distinctive transitions': people come to be industry analyst (1) after heavy industrial experience in IT field or (2) after a career in market research and technical publishing (Pollock and Williams 2016). The aim of the collaboration was for me to test this conjecture emerging from qualitative fieldwork by using a larger internet-generated digital data set.

Another conjecture emerging from interviews was that IT analysts consider establishing a one-person 'star-analyst' company as the pinnacle of their career: it offers comparable earnings and more autonomy than a job in the biggest companies. I wanted to seek confirmation whether once having joined top companies analysts tend to remain and exploit the reach offered by their position or prefer to trade their expertise independently, as the conjecture from the ethnographic study suggested.

To answer these questions, I entered a collaboration with a team of computer scientists who had previously developed a software capable of processing textual CV data. At the time, the idea was that we could parse IT analysts' CVs found on the web to yield information on the development of industry analysts' careers.

When returning to the field and mentioning this idea to my informants, I was made aware of the existence of an industrial database maintained by a US-based Analyst Relations firm that records career data of all known industry analysts, including the URL to their LinkedIn profile. With the help of text mining techniques, the team could then process the analyst career data from the industrial database, including the LinkedIn experience section, to extract one of more job elements referring to the different jobs the person held or is holding.

It was at that point that a PhD student at Edinburgh University as well as an IT analyst professional joined the team. Being familiar with the data set, he helped clarify the database structure and how it could be used to test the conjecture on 'distinctive transitions' deriving from the ethnographic study. Sorting analysts based on their 'power'—that is, how many times clients interact with them—and coverage, that is, what industry they analyse, he suggested to use the database to understand where the most successful analysts came from as well as how their careers pan out within the industry.

How Ethnography Affected Statistical and Computational *Modus Operandi*

Differently from many applied research ventures where the ethnographer is built into a pre-defined technological project to provide for lack of 'social factor' (Marcus 2002) or 'real world' problems (Goulden et al. 2017), in this project I had the chance to affect the conduct of the computational and statistical operations. In this section, I want to describe how my role influenced data processing.

As noted in careers literature, the ideal-typical career is the rarest of all (Rosenbaum 1979). Contrasting conjectures on the role of education (Bidwell and Briscoe 2010), professional background (Berger et al. 2016) and on-the- job training (Abbott 1988) seem to confirm that job histories in the IT sector are even more erratic. A high degree of variance has been indeed apparent in our data from the outset: with 464 company names and 1659 job titles from only the few hundreds of analyst profiles analysed for the initial pilot, it appeared there could be no way to otherwise identify patterns in career data but by drastic pre-processing. Given the variability in careers, this is an arbitrary but necessary step. Having conducted extensive ethnographic research, however, I could appraise every step of data processing through what I learnt from informants (i.e. IT analysts) during fieldwork.

In consultation with the IT analyst who joined the team, it was suggested that categorisation of companies would be based on size and seniority. Size would result from how many times the company was mentioned in the data sample. Seniority from how long ago a company was joined in the average. I expected that operationalising size in this way would confirm evidence from the previous ethnographic study of a few companies dominating the IT industry analyst market. It was found indeed that the three companies that were mentioned most frequently in the data where Gartner, Forrester and IDC (the three largest companies in the industry).

As well as with an interest for 'distinctive transitions', I came to this project with a view to extend the inquiry to a typology of industry analysts—those in smaller or one-person analyst houses—that are less likely to be reached at the industry events where I used to do most of my interviews. That would mean identifying which other companies occur later in a career, other than the largest analyst companies. Distinguishing by seniority would allow capturing which companies' analysts form or join when they want to trade their expertise outside the 'Big Three'. Three further categories of companies were therefore created (oldest, old, new) based on when in the career analysts joined them.

In addition to the ethnographer, the computer scientist and the domain expert, the team also included a researcher with expertise in life course events, including job-related events. He suggested the analysis of IT industry analyst careers to be carried out using the TramineR package. TraMineR is an R-package for mining and visualising sequences of categorical data (Gabadinho et al. 2011). Sequence analysis is a popular methodology in career studies (Abbott 1995; for a historical review see Rosenfeld 1992).

Using a sub-sample of random analysts as the reference group, the idea was to use sequence analysis to see if there was a pattern in top industry analysts' careers that was different in terms of provenance or career pathway from other typologies of analysts. In order to do so, a set of random analysts was collected to compare and contrast with top analysts (i.e. 'power' analysts as they were indicated in the database). For each profile, we exported job title, the company where the job was based and the date range of the job. By turning analyst profiles into job sequences, we could test the distinctive transition hypothesis (i.e. that the IT and the market research industries are the main feeders to the analyst industry) and find an answer to the question on the role of smaller analyst houses.

As the team approached the data with sequence analysis however, some difficulties emerged. The conjecture from the relatively limited set of interview data was indeed that the IT and the market research industries were the main feeders for the analyst industry. However, the data mined from analyst profiles showed that analysts have worked for so many different companies before taking up an analyst job. Taking away those who started their career directly from an analyst company, it was nearly one company per worker. Together with lack of authoritative lists of companies in feeder industries, it meant the team would struggle to find any meaningful result in the data as to where (which industry) people worked before taking an analyst job.

The resulting decision was to collapse all possible states concerning jobs in a non-analyst company into one single state (irrespective to company size or seniority).

However, as well as difficulties, the use of sequence analysis brought some unexpected revelations. By using sequence analysis, the team could now compare the evolution of IT industry analyst careers over a longer timeframe. For example, two plots were generated to contrast careers started after 2000 with careers started before 1990. The visual contrast showed a stark difference in the distribution of jobs over time.

While in careers starting before 1990 (see Fig. 4.1) the prevalent strategy was to access the analyst industry (yellow, green or orange sections of timeline) after having spent considerable time in other industries (purple sections of timeline), the trend seemed to have changed with careers beginning after 2000, with many more starting directly in the analyst industry (see Fig. 4.2). By showing that more and more analysts start their career directly in the industry, the use of sequence analysis defused concerns for distinctive transitions that seemed a rather central research theme when looking at interview data. Revelations such as this will be addressed in the next section where I will introduce the new notion of participative epistemology and attempt to re-calibrate relations between social science and data science.

Participative Epistemology

Through her work on epistemic cultures (Knorr Cetina 1999), Karin Knorr Cetina suggests that topical topics to investigate scientific knowledge production are (1) the epistemic subjects, that is, the authors of scientific research: the human collaboration as well as the digital platform producing data and how agency can rotate between them; (2) the objects of scientific research, and the reconfigurations through which the object gradually becomes amenable to the analysis; (3) the object-relation regime, which in the project under study means the interaction between the object of research (i.e. careers) and methodologies (ethnography, text mining and sequence analysis). In order to orient these epistemic categories to

Legend:
Purple = Jobs in non-analyst companies
Yellow = Jobs in smaller analyst houses
Green = jobs in Big Three analyst houses
Orange = jobs in one-man analyst houses
Grey = no data.

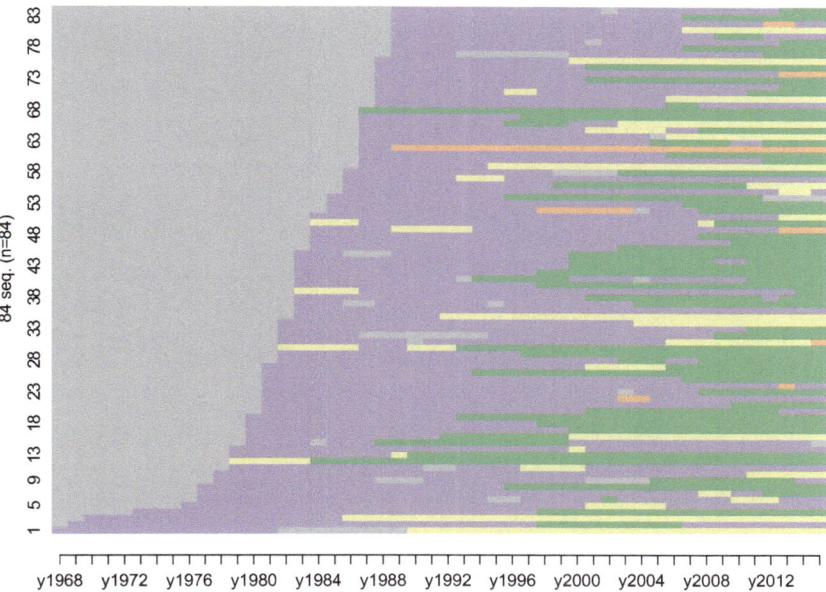

Fig. 4.1 This visualisation represents careers started before 1990. Horizontal axis represents time from beginning of career to time of the analysis. The vertical axis represents number of sequences. Careers are sorted from longer (no 1 at the bottom of the data region) to shorter (no 84 at the top). (Source: Author)

provide a situated account of social data science, I introduce the term participative epistemology. With participative epistemology I propose an empiricist notion of knowledge production whereby all those involved in the research endeavor (i.e. the epistemic subjects, the objects of scientific research and the object relation-regimes) are seen as both co-researchers, whose agency contributes to generating ideas, designing and managing the project, and drawing conclusions; and also co-subjects, participating in

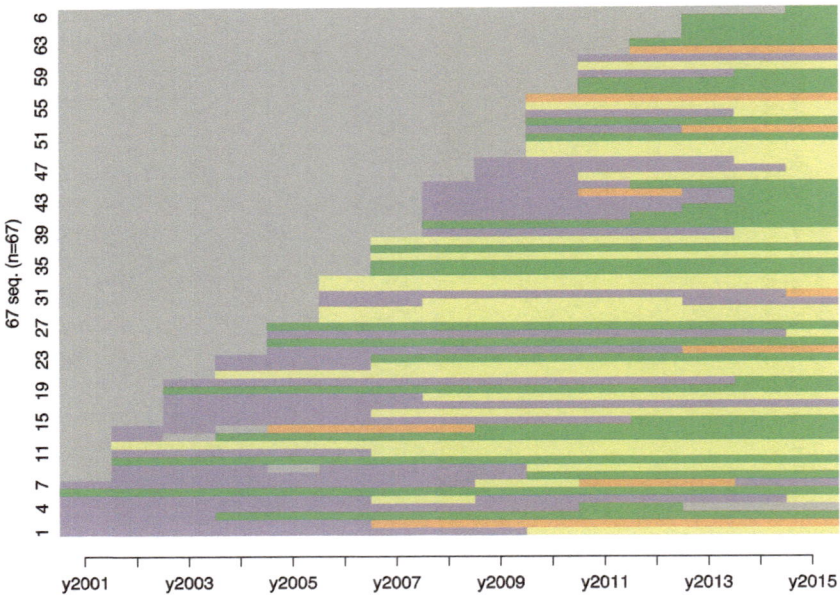

Legend:
Purple = Jobs in non-analyst companies
Yellow = Jobs in smaller analyst houses
Green = jobs in Big Three analyst houses
Orange = jobs in one-man analyst houses
Grey = no data.

Fig. 4.2 This visualisation represents careers started after 2000. Horizontal axis represents time from beginning of career to time of the analysis. The vertical axis represents number of sequences. Careers are sorted from longer (no 1 at the bottom of the data region) to shorter (no 67 at the top) to facilitate pattern identification. (Source: Author)

the activity that is being researched.[3] Incorporating an attention for quotidian aspects of knowledge production in social data science, the notion of participative epistemology extends existing notions[4] by looking closely at research subjects and their interactions. Additionally, it suggests a post-social (Knorr Cetina 1997) reading of knowledge production where participation is not limited to research subjects and extends to methods, their theoretical foundations (Collins 1984) as well as the research objects. This is not to be confused with the term participatory. The emphasis with participatory[5] is on an agenda that strives not only to know but to change the world, to help improve the living conditions of those in the workplace. Due to the more political participatory agenda, few projects have been realised with the ideological Other, that is, commercial firms. With participative epistemology, the emphasis is primarily on temporary joint epistemic work. Due to its post-social character, it does not regard shared value system as a necessary prerequisite of collaborative knowledge production.

In this section I show how the notion of participative epistemology became apparent in the project by describing how ethnography, text mining and sequence analysis as well as commercial data participated to shape the research object.

The initial assumption was that ethnographic insights could provide the initial conjectures to be tested: that industry analysts in their career go through distinctive transitions and that once having joined top companies analyst would trade their expertise independently. This could have configured the methodological connection as yet another case of a 'service' relationship whereby the social scientist can be seen as a sanitised species of 'problem owner', whose conjectures can be validated by the data scientist

---

[3] By choosing the term epistemology over ontology I do not mean to make a stand in the realism-constructionism debate and suggest that knowledge production practices are separate from the reality that they want to know (Daston 2000; Mol 2002). Rather, I comply with Latour's suggestion that no radical distinction between epistemology and ontology can be drawn (Latour 1999: 124).

[4] I refer here to concepts such as consilience (Wilson 1998), abduction (as in Kitchin 2014) and symphonic social science (Halford and Savage 2017) presented earlier in the chapter.

[5] This is not to be confused with 'participatory design' (Ehn 1988). Participatory design is an approach to design attempting to actively involve all stakeholders in the design process to help ensure the result meets their needs and is usable. In participatory design, participation is strictly intended to mean involvement of research users, that is, employees, partners, customers, citizens, end users in the research process.

through a mixture of assumption-free computational and statistical models. The contingencies of the project proved the scenario short-lived. As described in the above section, the conjectures originating from the ethnographic study had to be adapted—rather than simply tested—as a result of approaching them with a different method. This is one of the many evidences in our project of what Barad calls 'experimental entanglements' (Barad 2007: 148): knowledge comes after—and not before—awkward mixtures of methods and object–relation regimes. Ethnography and computational and sequence approaches—as well as the assumptions hardwired in commercial and professional networking data—are not just neutral means for testing some external (sociological) conjectures. They participate as theories to shape the research object. In the next section I show this first by discussing the role of assumptions embedded in commercial data. I will then focus on the epistemological convergences between the computational and the ethnographic approach. A third example of participative epistemology will be how sequence analysis helped reveal unconscious biases in the ethnographic approach.

Para-Ethnography

A particularly interesting aspect of participative epistemology in our project was the interaction of the researcher with 'commercial sociology' insights (Burrows and Gane 2006), represented in our case by the industrial data set and its 'power 100' metric. By commercial sociology Burrows and Gane (2006) mean innovations in research methods occurring within large commercial organisations becoming aware of the potential of data generated as a digital by-product of routine transactions between customers and business. Together with granting access to their database, the company curating analyst profiles provided the team with a ranking based on analysts' 'power'. This ranking was generated from transactional data on how many times analyst profiles were visited by users of their database. Being designed as a recommendation system the database excluded all those analysts who were not involved in client business. We found for example that Peter Sondergaard, the most prestigious analyst of the largest analyst firm in the industry,[6] is not in the list of 'power 100' for the simple reason that, as many other 'star' analysts, he does not operate client-business.

---

[6] Gartner is the largest Industry Analyst firm. Founded in 1979 by Gideon Gartner, the firm employs more than 4000 offices in over 80 countries and has a market value of $ 2.5 bn.

However, rather than dismissing this algorithm as a poor assessment of influence, commercial insights were used as an additional aid to inform our research strategy. To mitigate the client-business bias identified in the database, later in the project we came to consider job titles as well as the 'Power 100' algorithm as indicator of career outcome.

Domain expertise played an important role in revealing biases in the database and in suggesting how these could be mitigated. Borrowing the term from Marrow (1977) who used it to describe Kurt Lewin, Hoffman (2004) points out that practitioners collaborating to research are 'practical theorists' in that they help connect theoretical ideas with work on practical problems, thus bridging the gap between academic communities and the professional world. In our case, the practical theorist helped foreground the theories embedded in the structure of the commercial data set so that these could be compared at later stages with alternative theories linked with the use of computational and sequence methods.

The advent of applied research is generally seen in contrast to aspects of disciplinary epistemology that tend to be associated with a defence of academic autonomy. The danger for the ethnographer is to become too involved in the community under study (in my case industry analysts), especially when "natives" see value in research results. Losing objectivity and distance has been a well-rehearsed topic in the literature (O'Reilly 2009). Less common is the case, as with industry analysts, where natives are themselves cogent in studying each other in the very same terms as the ethnographer is (and producing research about it) or when indeed the informant, as in our case, lost his own distance to become an academic (i.e. by undertaking a PhD). In cases like this, where fortunate collaborations with the practical theorist allow delving deeper into the sorting algorithm, use of commercial data does not straightforwardly lead to a reduction of the autonomy or significance of empirical research. It can equally lead to new types of autonomy being created. Holmes and Marcus (2005) coin the term 'para-ethnography' for this situation, arguing that it entails a particular form of collaboration where expert informants become indeed 'epistemic partners' and their locally produced discourse can be taken to advance scholarly debate.

Marcus introduces another related concept of key importance: the para-site (Hadi and Marcus 2011). In the context of work among experts, doing theory should be seen in continuous relation to the distinctive ways of thinking apparent during fieldwork. The para-site is a kind of deliberate experimental disruption in the field research design with the intent of staging an encounter with the methods used by research partners. Marcus' concept of para-site aptly captures the idea in this book that social science encounters with data science are often nothing more than the result of the long-standing engagement with expert fields of practice where analytic methods are rapidly gaining currency.

COMPUTATIONAL METHODS

As with commercial expertise, the connections with other fields of academic expertise in this project were based on considering the computer scientist and the sequence analyst as research interlocutors. As evidenced with the concept participative epistemology, this does not mean that coordination rested on shared goals or intentions. There are indeed costs to taking participative epistemology seriously. As Barad (2007) insists, with coordinated labour it becomes methodologically fruitless in these experiments to delineate in advance distinct tasks, inputs and division of labour for 'data scientists' or 'social scientists'. In this section and the next, my attempt to separate methodological contributions is therefore purely for illustrative purposes.

In this project, the danger for the ethnographer to become too involved combined with the risk of losing direction in the interaction with affiliated ways of knowing, arguably raising the set of challenges to a level that Goulden et al. (2017) would define 'wild interdisciplinarity'. However, when observed in practice, the ethnographic and the computational approaches are divided by small differences in orientation and not by any significant epistemological chasm.[7] As Di Maggio puts it (Di Maggio 2015) when it comes to supervised machine learning, computer scientists are highly reliant on human judgement and call "golden standard" the human curated training set. When drastic decisions about curating corpora had to be taken such as in the case of company categorization, coordination between perspectives resulted in a preoccupation for these

---

[7] Ours and the projects discussed by Di Maggio have in common that the computational approach is utilized in projects including an ethnographic element. As noted in the remainder of the chapter, a similar observation would not necessarily apply when computational and ethnographic methods are applied individually in different projects.

decisions to be supported by strong empirical arguments without sacrificing computational advantages.

One aspect of this compromise is that the researcher claims ontological authority over which entities he will be researching on. This is to say that, after normalisation and categorisation, the raw data input for sequential analysis is actually created by the researcher. As arbitrary as it might sound, this is a necessary step for a number of reasons.

First because with nearly 500 company names and more that 1500 job titles for only a few hundred analyst profiles, categorisation was the only way to produce sequences with meaningful degrees of resemblance.[8] However, categories related to company size were controlled against evidence deriving from the previous ethnographic study: the outcome of the supervised model, for example, matched fieldwork evidence of few companies dominating the market.

Another reason to come to terms with the idea to alter the raw data using normalisation and categorisation at the cost of lengthy cross-checking exercises of data curation is that, being more inclined to rely on algorithmic solutions, the computer scientist would have let the data find the correlation.

Letting the data find the correlation, however, would have meant accepting the assumption that careers consist of fixed entities (job states) that have only one causal meaning that does not depend on the entity's location in the sequence. Due to evidence gathered from previous ethnographic study that a job in a small company can mean very different things if it is at the beginning or at the end of a career, I rejected the assumption of fixed entities and insisted to make the meaning of a given occurrence depend on its location within a sequence of events.

This was evident in the case of seniority categorisation of companies, based on when in the career people join them. A descriptive appreciation of career data of the top analyst sample evidenced that nearly all moves out of larger companies to establish a one-person company occurred in that

---

[8] Sequential analysis employs optimal matching techniques drawn from DNA analysis to cluster sequence patterns. It begins by establishing a distance between sequences in terms of the number of insertions, deletions and substitutions necessary to transform one sequence into another. By assigning varying costs to these insertions, deletions and substitutions the analyst can closely represent degrees of local resemblance.

subset. Although the evidence was not statistically significant, it was supported by observations deriving from the commercial database and it also matched the conjecture from the empirical fieldwork that practitioners consider establishing a one-person 'star-analyst' company as the pinnacle of a career.

Including complex human judgement in data pre-processing by creating a data ontology through classifying companies and ranking jobs according to fewer categories based on a combination of size and seniority also made sure that social sensibilities were written into these inscriptions, a condition that Bartlett et al. (2018) consider crucial to produce credible knowledge claims. The consequence of this was that the locus of legitimate interpretation was firmly fixed within training in the social understanding of IT analysts as opposed to being based on bringing to bear the tools and perspective of harder (computational and statistical) disciplines. This remained the case also at a stage where access to primary inscriptions has been shared to other members of the group.

SEQUENCE ANALYSIS

In the previous section I discussed the interaction between the computational and the ethnographic approach. In this section, I discuss how the sequence analysis approach helped reveal unconscious biases in the ethnographic approach. In doing so, I will distance myself from the view that statistics is a necessarily deterministic mode of thinking (Hacking 1975) devoid of any form of reflexivity and follow Collins (1984) by considering social statistics as a substantive theory. When I discuss assumptions embedded in sequence analysis, I refer in particular to the theoretical model common to statistical approaches that Collins (1984) characterises (referring to Keynes) as the 'atomic' model of the external world. Statistical methods assume that every event is casually connected with previous events belonging to its own series (a career) but it cannot be modified by contact with events belonging to another series (i.e. financial crises, personal circumstances, or other people's career moves). This assumption would not be acceptable from an ethnographic perspective.[9] However, as we insisted repeatedly throughout this chapter, commonality of epistemic modes is not a necessary requisite for participative epistemology.

---

[9] An insistence on the processual as a key feature of interactionist approaches can be found also in Housley and Smith (2017).

The previous section described how the ethnographic approach helped identify a categorisation of companies capable of including in the career model distinctions between repeated events (i.e. a job in a small company) that would be otherwise lost in a purely algorithmic search for statistically significant results. For example, our ethnographically informed decision to categorise companies by seniority (i.e. when people join them) as well as by size gave central importance to jobs in small companies later in a career as being the pinnacle of an analyst career.

However, more extended perspectives on the phenomenon such as those made possible by the use of sequence analysis also contributed to reveal biases ingrained in the ethnographic approach. By not assuming any privileged view on careers and by accepting for practical purposes to represent them as mechanical situations gave rise to the opportunity for new and different empirical accounts that helped reveal inconsistencies embedded in the retrospective character of the ethnographic approach.[10] For example, by using sequence analysis to compare careers over time, our original conjecture on the relevance of 'distinctive transitions' appeared to compete with other conjectures. It emerged that many early-career analysts start their career directly in the analyst industry. This means that looking at career moves between companies within the industry could reveal as much about outcome as our initial focus on 'distinctive transitions'.

Also, the assumption in ethnographic research is that knowing more is better.[11] While the statistical approach is about analysing data without preconceptions based on contextual factors emerging from antecedent studies. By looking at data from this different angle, we came to appreciate that not all individual sequences in our data set and not all points in a sequence were equally relevant for our question.

Combining ethnography with sequence analysis also helped bring a key practice in ethnographic approach such as the use of 'local informants' into further scrutiny. Science and technology studies (Latour 2005) already showed the informants are much more than mere translators or local brokers. In our project this particular combination of methods helped reveal that the ethnographic assumption that the more experienced an informant the more reliable their accounts are could give unjustified

---

[10] A similar point has been made by Evans and Aceves (2016) in the review of the use of natural language processing (NLP) in sociology when they say that 'unsupervised ML approaches can "discover" unnoticed, surprising regularities in these massive samples of text that may merit sociological consideration and theorization'.

[11] Until conceptual saturation is achieved (Glaser and Strauss 1967).

additional relevance to factors found in longer careers. Early results for example pointed out that successful analysts ('Power 100' analysts and CEOs) tend to have longer inter-professional careers.[12] Further investigation using sequence analysis methods, however, showed that without controlling our career data for age, or generation, we could not tell whether the correlation to outcome in this case was simply due to the overall length of the career as opposed to the inter-professional segment of the career more specifically.

CONCLUSION

It is apparent in our discussion that computational methods contribute to the ethnographic approach with a capacity to detect patterns related to less accessible informants. A similar finding was discussed with sequence analysis being able to show that career data gathered via ethnographic methods could be biased by dependency on specific gatekeepers.

The evidence deriving from this project goes against the prediction apparent in the literature that qualitative approaches can help identify case-specific idiosyncrasies (Leonardi 2011; Pentland and Feldman 2008). Quite on the contrary it seemed that in this particular project ethnographic approaches showed strengths in capturing the more visible patterns inferred from what was available at particular field sites, made apparent by conversations with informants that were, for reasons of expertise or seniority, more likely to interact with the fieldworker.

These results seems to resonate with Edwards and colleagues (Edwards et al. 2013) when they describe ethnography as an intensive strategy, as opposed to the more extensive, data science approach. The authors plot the distinctiveness of social media analysis in relation to more traditional research according to two dimensions: research design and research strategy. Social media analysis is said to offer an extensive research strategy, as opposed to qualitative methods (such as ethnography) that represent a more intensive strategy. The possibility to capture data in real time at the level of population and on an ongoing basis makes social media more processual than survey-based or experimental approaches (Sayer 1992).

---

[12] By inter-professional careers we mean careers that developed across different professions. In particular, we looked at the importance of career paths before taking up the first analyst job.

The other contribution of this chapter is to show methodological collaboration in action and in its local interaction with the research object as opposed to by conceptual binaries (i.e. micro/macro, intensive/extensive, etc.).

In our specific case, a focus on participative epistemology suggests avoiding the temptation to unify social data science. In contrast, it claims to only apply to the local encounter of ethnography, computational and statistical methods with a particular type of phenomenon (careers) and a particular form of (commercial) data. More in general, the concept of participative epistemology invites to make sense of the multiplicity of epistemic forms and recognise the extent to which our understanding of the social always evolves together with new combinations of phenomena, tools, techniques and narratives about disciplinary affinities.

References

Abbott, A. (1988). *The System of Professions. An Essay on the Division of Expert Labor*. Chicago: The University of Chicago Press.

Abbott, A. (1992). From Causes to Events: Notes of Narrative Positivism. *Sociological Methods & Research, 20*(4), 428–455.

Abbott, A. (1995). Sequence Analysis: New Methods for Old Ideas. *Annual Review of Sociology, 21*, 93–113.

Barad, K. (2007). *Meeting the Universe Halfway: Quantum Physics and the Entanglement of Matter and Meaning*. Durham, NC: Duke University Press.

Barry, A., & Born, G. (2014). *Interdisciplinarity: Reconfigurations of the Social and Natural Sciences*. London: Routledge.

Bartlett, A., Lewis, J., Reyes-Galindo, L., & Stephens, N. (2018). The Locus of Legitimate Interpretation in Big Data Sciences: Lessons for Computational Social Science from -Omic Biology and High-Energy Physics. *Big Data & Society, 5*(1), 1–15.

Bateson, G. (2002 [1979]). *Mind and Nature: A Necessary Unity*. Cresskill, NJ: Hampton Press.

Berger, G., Gan, L., & Fritzler, A. (2016). *How to Become and Executive*. Retrieved from https://www.linkedin.com/pulse/how-become-executive-guy-berger-ph-d-?published=t.

Bidwell, M., & Briscoe, F. (2010). The Dynamics of Interorganizational Careers. *Organization Science, 21*(5), 1034–1053.

Blok, A., & Pedersen, M. A. (2014). Complementary Social Science? Quali-Quantitative Experiments in a Big Data World. *Big Data & Society, 1*(2), 1–6.

Bohr N (1957 [1938]) Atomfysik og menneskelig erkendelse. Copenhagen: Schultz Forlag.

Boyer, D. (2014). Reflexivity Reloaded: From Anthropology to Intellectuals Critique of Method to Studying Sideways. In T. H. Eriksen et al. (Eds.), *Anthropology Now and Next : Essays in Honor of Ulf Hannerz*. New York: Berghahn Books, Incorporated.

Burrows, R., & Gane, N. (2006). Geodemographics, Software and Class. *Sociology, 40*(5), 793.

Collins, R. (1984). Statistics Versus Words. *Sociological Theory, 2*, 329–362.

Daston, L. (2000). The Coming into Being of Scientific Objects. In L. Daston (Ed.), *Biographies of Scientific Objects* (pp. 1–14). Chicago, IL: Chicago University Press.

Di Maggio, P. (2015). Adapting Computational Text Analysis to Social Science (and Vice Versa). *Big Data & Society, 2*(2), 1–5.

Edwards, A., Housley, W., Williams, M., Sloan, L., & Williams, M. (2013). Digital Social Research, Social Media and the Sociological Imagination: Surrogacy, Augmentation and Re-orientation. *International Journal of Social Research Methodology, 24*, 313–343.

Ehn, P. (1988). *Work-Oriented Design of Computer Artifacts*. PhD diss., Arbetslivscentrum.

Evans, J. A., & Aceves, P. (2016). Machine Translation: Mining Text for Social Theory. *Annual Review Sociology, 42*, 21–50.

Gabadinho, A., Ritschard, G., Mueller, N. S., & Studer, M. (2011). Analyzing and Visualizing State Sequences in R with TraMineR. *Journal of Statistical Software, 40*(4), 1–37.

George, G., Haas, M., & Pentland, A. S. (2014). Big Data and Management. *Academy of Management Journal, 57*(2), 321–326.

Glaser, B. G., & Strauss, A. L. (1967). *The Discovery of Grounded Theory: Strategies for Qualitative Research*. Chicago: Aldine Pub. Co.

Goulden, M., Greiffenhagen, C., Crowcroft, J., McAuley, D., Mortier, R., Radenkovic, M., & Sathiasaleen, A. (2017). Wild Interdisciplinarity: Ethnography and Computer Science. *International Journal of Social Research Methodology, 20*(2), 137–150.

Hacking, I. (1975). The Emergence of Probability. Cambridge: Cambridge University Press.

Hadi, D., & Marcus, G. E. (2011). In the Green Room: An Experiment in Ethnographic Method at the WTO. *PoLAR, 34*(1), 51–76.

Halford, S., & Savage, M. (2017). Speaking Sociologically with Big Data: Symphonic Social Science and the Future for Big Data Research. *Sociology, 51*(6), 1–18.

Healy, K., & Moody, J. (2014). Data Visualization in Sociology. *Annual Review Sociology, 40*, 105–128.

Hoffman, A. (2004). Reconsidering the Role of the Practical Theorist: On (re) Connecting Theory to Practice in Organization Theory. *Strategic Organization, 2*(2), 213–222.

Holmes, D., & Marcus, G. (2005). Cultures of Expertise and the Management of Globalisation: Towards the re-functioning of Ethnography. In A. Ong & S. Collier (Eds.), *Global Assemblages: Technology, Politics and Ethics as Anthropological Problems* (pp. 235–252). Malden, MA: Blackwell.

Housley, W., & Smith, R. J. (2017). Interactionism and Digital Society. Qualitative Research, 17(2), 187–201.

Kitchin, R. (2014). Big Data, New Epistemologies and Paradigm Shifts. *Big Data & Society, 1*(1), 1–12.

Knorr Cetina, K. (1997). Sociality with Objects: Social Relations in Postsocial Knowledge Societies. *Theory, Culture & Society, 14*(4), 1–30.

Latour, B. (1999). *Pandora's Hope: Essays on the Reality of Science Studies.* Cambridge, MA: Harward University Press.

Latour, B. (2005). *Reassembling the Social: An Introduction to Actor-Network-Theory.* Oxford: Oxford University Press.

Leonardi, P. M. (2011). When Flexible Routines Meet Flexible Technologies: Affordance, Constraint, and the Imbrication of Human and Material Agencies. *MIS Quarterly, 35*(1), 147–167.

Marcus, G. (2002). Intimate Strangers: The Dynamics of (Non) Relationship Between the Natural and Human Sciences in the Contemporary U.S. University. *Anthropological Quarterly, 75*(4), 519–526.

Marres, N., & Weltevrede, E. (2015). Scraping the Social? Issues in Real-Time Social Research. In J. Bouchard, É. Candel, H. Cardy, & G. Gomez-Mejia (Eds.), *La Médiatisation de l'Évaluation.* Berlin: Peter Lang.

Marrow, A. (1977). *The Practical Theorist: The Life and Work of Kurt Lewin.* New York: Teachers College Press.

Mol, A. (2002). *The Body Multiple: Ontology in Medical Practice.* Durham, NC: Duke University Press.

O'Reilly, K. (2009). *Key Concepts in Ethnography.* London: Sage.

Pentland, B. T., & Feldman, M. S. (2008). Designing Routines: On the Folly of Designing Artifacts, While Hoping for Patterns of Action. *Information and Organization, 18*(4), 235–250.

Pollock, N., & Williams, R. (2016). *How Industry Analysts Shape the Digital Future.* Oxford: Oxford University Press.

Radin, P. (1927). *Primitive Man as Philosopher.* New York: D. Appleton and Company.

Randall, D., Harper, R., & Rouncefield, M. (2005). Fieldwork, Ethnography and Design: A Perspective from CSCW. In K. Anderson & T. Lovejoy (Eds.), *EPIC 2005: Ethnographic Praxis in Industry Conference* (pp. 88–99). Seattle, WA and Arlington, VA: Redmond, American Anthropological Association.

Rosenbaum, James E. (1979). "Organizational career mobility: promotion chances in a corporation during periods of growth and contraction." American Journal of Sociology, 85(1): 21–48.

Rosenfeld, RA. (1992). Job mobility and career processes. Annual Review of Sociology, 18: 39–61.

Sandvig, C., & Hargittai, E. (2015). How to Think about Digital Research. In E. Hargittai & C. Sandvig (Eds.), *Digital Research Confidential: The Secrets of Studying Behavior Online*. Cambridge, MA: MIT Press.

Savage, M. (2015). Sociology and the Digital Challenge. In P. Halfpenny & R. Procter (Eds.), Innovations in Digital Research Methods. London: Sage.

Sayer, A. (1992). *Method in Social Science: A Realist Approach* (2nd ed.). London: Routledge.

Wilson, E. (1998). *Consilience:The Unity of Knowledge*. New York: Knopf.

CHAPTER 5

Ethnography as Data Science

Abstract This final chapter discusses ethnography as a core discipline in data science. Taking the symbolic AI versus connectionism debate as a starting point, it refers to the ethnomethodological notion of accountability to show how ethnography can contribute to the debate around making machine learning more interpretable and explainable. It observes that the main focus of social studies of data science has been in the area of data curation and it argues to expand the scope of the "algorithmic drama" to include the architectural aspects of machine learning. It states that given the direction deep machine learning is taking towards the design of 'ensembles' of networks and decision trees, social scientists should devote more attention to how algorithms interact.

Keywords Explainable AI • xAI • Interpretability • Accountability of algorithms

In Chap. 4 I showed that the coming together of ethnographic, computational and statistical methods under the umbrella of social data science can be seen as contributing empirical sociology with new sensitizing concepts, inviting empirical sociology to work side-by-side with other epistemic modes. Showing how the machine learning artillery can extend the methodological (as well as theoretical) toolbox of ethnography, however, only responds to part of the argument of this book. The other complementary

G. M. Campagnolo, *Social Data Science Xennials*, https://doi.org/10.1007/978-3-030-60358-8_5

aspect is to show the extent to which ethnography can be considered as a core discipline in data science. To conclude the exploration, this chapter suggests different politics to understand contemporary interdisciplinary forms and the logics of ethnography involvement in data science. While the previous chapter considered data science and ethnography as the bearers of different epistemic modes, this one takes ethnography and data science as sharing the same knowledge objective: to reveal the methods of production of machine learning. Referring to the ethnomethodological notion of accountability, it will discuss potential future directions for how phenomenological sociology inspired interactionist studies can contribute to the debate around making data science more interpretable and explainable.

The Socio-Materiality of Algorithms

Existing research in the field of Science and Technology Studies (STS) offers an important case to develop a more nuanced and grounded understanding of the constitution of algorithms (Muniesa 2011; Introna 2016; Neyland 2016). One important effort in this direction is, for example, the work of Jaton (2017) on the constitution of image-processing algorithms for salience-detection. Reacting against the depiction of algorithms as inscrutable black boxes, the article successfully focuses on the role of ground truth and the nature of the data that informs algorithms.

There are, however, many other aspects of the inner mechanisms of machine learning that haven't been yet explored sociologically. Algorithms are uncertain configurations for different reasons, not only for the pre-processing that goes in the data that informs the classifier. For example, although he recognises that data processing (i.e. feature engineering in machine learning terms) is one of its most time-consuming aspects, Domingos (2012) reminds us that there are many other subtleties that need to be taken into consideration in designing a machine learning project. These include: which algorithm to use, how to measure its scoring ability in the classification task, how to choose among high-scoring classifiers and how to combine these three aspects.

So far it seems that the main focus of STS detailed investigations of data science has been in the area of feature engineering, that is, how to represent the input, looking at the laboratory practices through which ad-hoc data calibration is achieved. Also influenced by theoretical traditions in which materiality is prominent, such as 'media materialism' (Parrika

2015), this chapter claims that the role of ethnography is to expand the social study of data science to cover further areas, including the architectural aspects of machine learning.

Performing machine learning is not solely about doing it correctly. There are so many more occasions in which deep machine learning is utilised which have nothing to do with producing or working out any new algorithm. Instead, these are situations that mobilise algorithms produced earlier or by others in conversations regarding for example areas of application. The ability of AI scholars and practitioners to use machine learning in social situations depends on their understanding of the circulation, provision, and reception of their algorithms by others.

This is especially true for machine learning where re-embedding is constitutive of its very nature. As described by Cardon et al. (2018) algorithms are designed to produce outputs for the only purpose of maximising the predictive performance of the programme. Taking them to mean something in any other area of human activity is to de-contextualise them. As phenomenological sociology points out, meaning making is intrinsically indexical (Garfinkel 1967). Similarly to the inner functioning of the calculator in the connectionist approach, the information produced by algorithms does not come in association with any given meaning. It is the context in which the output is put to use that decides what meaning to assign to it. Ethnography can lend data science the tools to attend to this meaning-making process.

ETHNOGRAPHY AND ARTIFICIAL INTELLIGENCE

As mentioned in Chap. 1, ethnomethodology-informed ethnography had an extended affair with symbolic artificial intelligence as demonstrated by the commentary of Lucy Suchman's work made by Herbert Simon, one of the founders (with Alonso Vera) of the symbolic approach (Suchman 1987; Vera and Simon 1993) and later towards the declining moment of the symbolic approach by the Collins–Dreyfus conversations on expert systems (Collins 1990; Dreyfus 1992). At the time, the discussion on AI was a discussion on human intelligence. The claim AI had about replicating human intelligence was an important one as it offered a common ground for multiple disciplines to come together. The situation has changed since.

The new wave of connectionism being marginalised for so many years shows little engagement with other disciplines, let alone social science.

What made it easier for social scientists to engage was the strong claim made by symbolic AI that they wanted to create an intelligent machine. Social scientists could confront AI scholars over a definition of cognition. What Lucy Suchman has been battling with since her work at Xerox and her 'Plans and Situated Action' book of 1987 is that the idea of the definition of logical rules in AI was completely devoid of the corporeal, situated, implicit, embodied, collective and contextual forms of the perception, orientation and decisions of human action (Suchman 1987). If one reads the debate again (say, e.g. Collins 1990), the claims sociologists of knowledge made against AI back in the late 1980s/early 1990s are now largely outdated.

The proponents of neural networks these days do not make any more general claims about intelligent machines. The data are less and less structured. The problems addressed are larger but less and less expert.[1] The fact that there are no strong claims for these systems to produce a generalised set of rules and abstractions that can be used beyond the local context means that there seems to be less of a hook for social scientists to critically connect.

Also, a new population of data scientists, hackers and do-it-yourselfers entered the previously very closed (and academic) field of AI. While on the one hand they have computer development skills that weren't previously required, on the other hand, their epistemological orientation is akin to what Domingos (2012) defines as "folk knowledge", lacking the vocabulary of the connectionism/constructivism debate that so clearly linked earlier AI proponents with conversation across disciplines.

This 'new' AI makes claims of distinguishing itself from academic AI and that the data they are using is messy, noisy real-world data—as opposed to those 'toy' data sets used in laboratories (Bengio et al. 2013). In erecting these distinctions against the earlier proponents of knowledge engineering, the new connectionist programme portrayed so well by Cardon et al. (2018) in their article 'Neurons Spike Back' also ditches the check of academic accountability in the form of a request to demonstrate what is the relevance of the AI work for the progress of scientific knowledge apart from that of solving specific business problems.

---

[1] It is not by chance that the resurgence of connectionism is initially linked to problems such as spam detection, collaborative filtering techniques for making recommendations, inventory prediction, information searches, or the analysis of social networks.

Given all this, what makes me think that phenomenological sociology, ethnomethodology and interactionism can still candidate to befriend AI and contribute to co-construct the typical history of compromises that the interdisciplinary debate brings with it? To answer this question, I will now look at the area of the AI debate around interpretability and how I think it can benefit from sociological insights. Before that, it is important to assess the claims of theoretical neutrality made by the new connectionism sketched earlier in this section.

THE MODERN PROMETHEUS

It is apparent even to its own proponents that data science is at a point where a study of its own ways of working is desperately needed. As famously observed by Breiman (2011)—a classical statistician turned data scientist—there seems to be in machine learning a trade-off between interpretability and accuracy. Traditional statistical methods such as linear regression provide a picture that can be understood by many outside the AI field of the relation between response and predictor variable. However, statistical methods are less accurate when it comes to identifying a good predictor. AI methods such as neural networks or random forests do not seem to provide a clear picture of the relation between response and predictor variable but appear to be much more accurate in identifying a good predictor. With respect to the interpretability of AI methods, it is often stated that a unique feature of new intelligent machines is that the same scientists who have created artificial intelligent agents cannot fully anticipate their behaviours and that some of their attributes are impossible to formalise analytically. This lack of understanding has been identified as the 'AI Knowledge Gap': the number of unique AI systems grows faster than the number of studies that characterize these systems' behaviour (Epstein et al. 2018) and it is suggested to be given to the interdisciplinary scholar to fill.

At present, the scientists who study the behaviours of these agents—in what has been named explainable AI or xAI—are predominantly the same scientists who have created the agents themselves. The results of engaging in evaluation activities called optimisation or meta-learning is more deep learning (Voosen 2017). But methodologies aimed at maximised algorithmic performance are not optimal for conducting scientific observation of the properties and behaviours of AI agents.

Even the leaders in the AI field recognise that different investigative approaches are needed. For example, Epstein et al. (2018) suggest the Turing Box. On one side, AI contributors upload existing and novel algorithms to be studied scientifically by others. On the other side, AI examiners develop and post machine intelligence tasks designed to evaluate and characterise algorithmic behaviour.

When it is not a showcase for algorithm performance optimisation, the narrative around explainability in AI is heavily informed by behaviouristic metaphors. For example, the recent manifesto on 'Machine Behaviour' claims that the study of intelligent machines can be conducted empirically by conceiving intelligent machines not as engineering artefacts, but as a class of actors with particular behavioural patterns and ecology (Rahwan et al. 2019). However, Rahwan et al. (2019) begin their paper by saying that they want to establish a field 'akin to how ethology and behavioural ecology study animal behaviour'. It is ironic that while promising to build machines that replicate human intelligence, when it comes to propose a scientific study of machines empirically, the preferred approach seems to be inspired by ethology. It appears that AI scholars see the fact that deep learning machines achieve cooperation and social order without knowing each other's inner mental states to be at odds with a sociological explanation. There is, however, no need to turn to the study of animal behaviour to account for spontaneous organisation of collective activities, that is, activities that take place at a level that is at loss of single individual capabilities. In the remainder of this chapter I will show that there are approaches in the field of phenomenological sociology that can deal with patterns of complex forms of human interactions inductively and without assuming mental states.

Seeing Work Where Others See Facts

Interactionist approaches and affiliated ethnographic methods can be an important candidate to help make data science more interpretable also for another reason. With their deep reflexive concern with the theatre of its own operations, they can reveal how algorithms are in fact accomplished and co-created through a series of mundane and everyday actions.

The AI narrative around fear of the potential loss of human oversight over intelligent machines may be seen as reproducing a methodological disengagement that tends to obscure the intentionality and the responsibility that the algorithm designers have when they select and train the algorithm, when they decide which problems can be addressed algorithmically and which cannot, when they recognise or refuse to recognise the ideational and institutional matrix in which algorithm performance is embedded.

Machine learning is presented even by the most knowledgeable proponents as being naively empiricist. Nature, it is said, forms the output y from the inputs x by means of a black box with complex and unknown interior. Current accurate prediction methods are 'natural' in that they also are complex black boxes and even slightly less inscrutable than nature. However, on closer inspection this does not seem to be the case.

To win, the algorithmic approach requires the broadest and most elementary digital data possible. As an input to the calculator, the composition of the world has to undergo a profound shift in terms of atomisation and granularisation. While the 'toy' and expert knowledge worlds of symbolic machines consisted of small, limited worlds that had been cleaned up and domesticated via a framework of intelligible and interdependent features, connectionist machines operate in a world in which the data must be as atomised as possible in order to deprive it of any explicit structure. Even if the data contain regularities, compositional relationships, global styles, etc., these elements must be highlighted by the calculator and not by the programmer. Therefore, data are not made available to the perception of calculators in a 'raw' and 'immediate' form, but rather are subject to atomisation and dissociation in order to transform it into the most elementary possible standardised digital signs.

To create these inputs, a new metrology of sensors, recordings and databases constitutes an essential infrastructure for transforming images, sounds, movements, clicks, or variables of all types into these giant vectors required by connectionist machines (MacKenzie 2017). For example in a text, to input a word into a neural network it must be embedded into a vectorial space that measures its distance from the other words in the corpus. In other words, the designers of connectionist machines still carry out artificial operations to transform data into another representation system.

The claim for algorithmic naturalness is associated with the continued use of a realist perspective whereby algorithms exist to one side and society to the other. For each of them, there is a thing called social science that addresses itself to one kind of object; and there is a thing called data science that addresses itself to another. The only controversy is about whether current flirtations between the two should be consummated. This debate thus operates entirely within an unquestioned, shared space, which Fitzgerald and Callard (2015) call 'the regime of the inter-'.

Studying algorithms from a truly sociological perspective implies abandoning the realist stance that postulates the occurrence of algorithms and society as existing in two separate domains working according to their own autonomous systems of rules and that only interact mechanically through their outputs. It means reversing the perspective, adopting an anti-essentialist and non-individualistic epistemological stance, assuming that algorithmic agents and social actors not only interact with each other but that they are one another or that they own one another (Latour et al. 2012) in an interactive process that in turn produces roles or identities as an achievement.

Accountable AI

As J. Markoff (2015) emphasises, the entire history of computer science is underpinned by an opposition between people promoting intelligence in machines (artificial intelligence—AI) and epitomised by the obsession with robotics; and people looking to distribute intelligence between humans and machine interfaces (intelligence amplification—IA), which would give rise to the human–computer interaction (HCI) school (Grudin 2009). The topic of this chapter is how to distribute agency again between science and society and put ethnographers in a position similar to that they had in relation with symbolic AI. As mentioned earlier, to do so I identified the explainability/interpretability debate as a promising domain. In this section, I look at the existing status of the debate within the AI discipline before suggesting an ethnography inspired approach based on the ethnomethodological notion of accountability.

One typical context where explainability concerns emerge is the discussion around AI bias. Take for example job recruitment. A firm can employ more men as its predominantly male management has hired fewer women due to unconscious bias. If that company uses employment data to train an IA to assess job applications, the lack of data on women will make it harder for the system to judge their aptitude with confidence, putting them at a disadvantage. At the recent virtual conference on Machine Learning (ICML), the answer given to this typical problem by Sanghamitra Dutta at Carnegie Mellon University is to create an 'artificial' but 'fairer' data set to train the algorithm. The problem is that if you develop an 'ideal' data set using mathematical techniques to equalise the amount of information on each group, predictions become less accurate, that is, the next hire will not resemble the hiring pattern of that organisation which was the original target of the classifier given the existing training set. However, although less accurate, this approach gives statistical guarantee of fairness.

Apart from raising concerns as to whether fairness can be automated (Watcher et al. 2020), it is notable how the solution proposed here resembles a hybrid symbolic/connectionist approach in that it resurfaces the idea of a 'toy' data set that was heavily criticised in the scenario discussed earlier in this chapter.

Despite this shortcoming, the effort of rushing the AI discipline into defining some of its terms is promising. It provides a hook for multiple disciplines to concur into discussing important aspects of machine learning. However, instead of a *déjà vu* of old-fashioned IA ideology that might run into the same criticisms of the representational approach, a phenomenological sociology inspired approach to explainable IA suggests to look more closely to the inner mechanisms of the algorithm.

As opposed to looking at social science as a resource aspect in the process, one good just to evaluate AI and the implications of algorithms after the fact, I want to discuss how sociological inquiry can take a foundational place in the very notion of algorithm architecture design conceived as a distributed assemblage of human and non-human social actors.

If the recognition is that what should be done is to find a language to describe mutualistic symbioses of society and algorithms—that is, how the artificial community and the social community interact—why not to try and study the dynamics of algorithms using social theories?

In order to define how this approach would look like, I refer back to the 'technomethodology' programme (Dourish and Button 1998), which in the eyes of the authors represents a fusion of the foundational principles of technology design on one side and ethnomethodology on the other. In that work, the authors make reference to ethnomethodology's notion of accountability which suggests that 'the activities whereby members produce and manage settings of organised everyday affairs are identical with members' procedures for making those setting 'account-able' (Garfinkel 1967: 1). By referring to further applications of this notion to workplace studies (Heath and Button 2002), they suggest that the emphasis of accountability can extend to how technologies (records, guidelines, categories) come to participate in producing the accountable order of work practices.

Similarly to the technomethodology approach and in line with a Human–Computer Interaction view on machine learning, I want to explore the notion of accountability to suggest that the mechanisms that provide a context to make system activity rationalisable are within the circumstances of its activity rather than outside of it (as in a manual or a testimony). Similarly to what Dourish and Button suggested, to widen the participation to and support the public understanding of data science I claim that a field strategy is needed to bring to the fore more meaningful accounts of how algorithms go about performing their activities.

Other scholars in Science and Technology Studies have attempted to apply the concept of accountability to the study of algorithms. One important reference is the pioneering work of Daniel Neyland (Neyland 2016). Neyland studied algorithms for video-based surveillance in airports. The 'algorithms' discussed by Neyland address a fully developed software package using a Codebook algorithm for object detection that could be 'dropped into' existing surveillance architectures to monitor the airport space and send alerts to system operators. The case described in Neyland's paper has commonalities with the traditional system design setting where computer scientists consult domain experts, in this case surveillance operators. Looking more specifically, the type of algorithm discussed in Neyland's paper is one that sifts through the CCTV feed to identify relevant events according to a rule-based programme. The surveillance operator would subsequently tag the selected video as relevant or irrelevant. The learning algorithm used to generate probabilistic routes reconstruction in connection of abandoned luggage uses probabilistic trees. In the case described by Neyland, at all times the system could be made entirely transparent to the domain experts.

In order to fit the requirements of the research agenda proposed in this chapter, the approach to algorithmic accountability exemplified by Neyland should be extended in two directions. The first is to apply it to the case of deep learning algorithms, where visual inspection of the inner decisions of the system is extremely unlikely. The second is to conceive an application of the notion of accountability to the practices of algorithm design and testing. In studies on accountability one important question is 'accountability for whom?' In Neyland's case, the chain of algorithmic accountability included an ethical committee to whom the researcher had to report.[2] Critically, the work of the ethnographer was not explicitly one to contribute to the design of the system.

My aim is to understand how the moment-to-moment sense-making ethnomethodologists are interested in can be productively interrelated within the constituted orders of accountability (Suchman et al. 2002:164) that are discussed within the contemporary AI debate on interpretability (Diakopoulos 2016). Let us understand how this intersection can be articulated.

ACCOUNTABILITY AND EXPLAINABLE AI

According to Lipton (2018), the two broad aims of work on interpretability in machine learning are transparency and post hoc explanation. Transparency addresses how a model functions internally, whereas post-hoc interpretation concerns how the model behaviour is reported (Lipton 2018). While the larger set of existing technique fall into the category of post hoc explanation, the techniques of AI interpretability that more closely resonate with the ethnomethodological notion of accountability are those that deal with the production of accounts on how the model work as part of the process—those techniques that Lipton lists under the 'transparency' label.

This is because in ethnomethodological writing, accountability is an innate feature of social action whereby members of a society do entertain the making visible and reportable of an action to an audience as part of the very essence of any social action. As mentioned earlier regarding the technomethodology programme, an important part of the ethnomethodological notion is that the account is inseparable from the action and it can only emerge as part of that action. The postulate of this definition is that, being inseparable from human conduct, accounts are produced by those who undertake that action in sync with the undertaking of that action. There are no good or bad accounts of social action and as long as something is considered social there is no delay or opt out from it being accountable.

We now return to the AI register and look more in detail to how the notion of accountability can contribute to the xAI debate. Transparency, Lipton says (2018), can be achieved at different levels: at the level of the entire model, at the level of the individual components such as parameters

---

[2] As the author himself concedes, Neyland's study can indeed be also seen as an ethnographic study of ethics.

and at the level of the training algorithm(s). In the most literal sense, a model might be called transparent if a person can contemplate the entire model at once.[3] This literal understanding of transparency generates a debate around size of the model and the time that would take for a human to take the input data together with the parameters and step through every calculation. The typical conclusions are that interpretable models are those that can be "readily presented to the user with visual or textual artifacts" (Ribeiro et al. 2016). However, as discussed in Chap. 2 when we introduced the notion of appresentation (Schutz 1971), social actors are notoriously able to see more than concretely presented perspectival details. A degree of ignorance is constitutive of successful sense making (McGoey 2012). It is ordinary experience of sense-making that what is not concretely 'present' is implicated along with what is immediately apparent and, reciprocally, it determines the sense of what is seen (a house with an interior, etc.). In other words, a phenomenological take on the xAI debate does not assume that the way to making it accountable is the full disclosure of the internal functioning of a machine learning model. Quite on the contrary, as mentioned in Chap. 2 when discussing the ethnomethodological notion of vagueness, the orderliness of a situation is hampered if the interactant seeks excessive explanation. In phenomenological terms, accountability does not mean that users have to have a complete understanding of how results are produced.

What accountability really means is that, if you look close enough, design decisions will appear as being taken in ways that safeguard participants' interests and affect participants' abilities for action selectively locally and in future occasions. For an ethnographer, these more mundane parts of machine learning are as important as and indistinguishable from the mathematical calibration of the parameter in a classifier for the development of an algorithm.

For example, talking about a real-world example analysing the speed of trials based on court data systems, Breiman (2011: 207) reports that when using decision trees—a highly interpretable form of machine learning—results are immediately apparent to the non-expert audience (in his case an assembly of judges). The accountability of the decision trees is in the fact

---

[3] According to AI, transparency can also be understood also in more discrete terms. It can for example refer to how each part of the model, that is, the input themselves, each node in a decision tree (e.g. making it correspond to a textual description) and the shape of the error surface admit an intuitive explanation.

that the assembly immediately starts to comment algorithmic results saying: 'I knew those guys were dragging their feet'. In another note from the same paper when discussing another more inscrutable machine learning method called random forests, Breiman confirms that in his experience despite successes in prediction in the medical sector, given that doctors can interpret logistic regressions and not 50 decision trees hooked together in random forests, in a choice between accuracy and interpretability the medical professionals will always go for interpretability.[4] However, if it is true that machine learning is finding its way in so many areas of our everyday life from search engines to personalisation systems, from the running of political campaigns to designing urban spaces, there should be ways in which people socialise them and attach a sort of shared meaning to them. This is the task for the ethnographer of data science.

If the main ethnomethodological intuition is correct and accountability is an irreparable feature of social order, then phenomenological sociology can contribute to the xAI debate a way to overcome the Breiman argument of the 'two cultures' in data science (Breiman 2011) by offering a way to show that deep machine learning models are as witness-able as the logistic regressions and the decision trees mentioned in the earlier examples.

How Algorithms Interact

In order to grasp more analytically how an accountability-informed field strategy to study deep machine learning would look like, I want to refer to Erving Goffman, a social scientist widely cited by sociologists who study some of the most inscrutable algorithms in business: financial algorithms (Knorr Cetina 2009; MacKenzie 2019). Goffman says that social situations 'begin when mutual monitoring occurs' (Goffman 1963: 18). Any

---

[4] Statements such as those found in Lipton (2018: 43) that in some cases interpretability may be at odds with the broader objectives of IA are highly problematic. Moreover, the hubris of the argument in that paper goes as far as maintaining that 'the short-term goal of building trust with doctors by developing transparent models might clash with the longer-term goals of improving health care' as if (1) an improved health care would be possible without doctors being able to interpret the functioning of their decision-support algorithms and (2) the longer-term goals of improving health care are indeed decided by machine learning experts. There are no good or bad reasons for providing accountability. Considerations regarding institutional 'biases' against new methods are constitutive of sense making as much as predictive power.

behavioural unit (human or machine) whose activity draws the attention of another (human or machine) into a process of mutual monitoring becomes participant of a social situation.

According to Goffman, social situations are subject to ordering by frames which establish the meaning of what is going on and regulate participants' activities (Goffman 1974: 10–11). Similarly to how Vollmer (2007) used concepts from Goffman's frame analysis to study financial numbers, ethnographers of data science can investigate the conditions under which machine learning is accountable by revealing how machine learning is 'framed' and 'keyed' in social situations.

There are at least two distinctive ways of framing the utilisation of machine learning models. One is the actual reproduction of machine learning models—which is the frame the xAI debate refers to. This is the realm of the calculation, where numbers, computer code and mathematical formulations are produced. All that matters in this frame is the measure of the scoring ability of an algorithm in a classification task. In this framing, which Vollmer called reproductive (Vollmer 2007: 583), the reference of the inner workings of the algorithm to any reality external to the set of signs and operations defined by computational practices is bracketed.

Still, this is only one part of the story. As Domingos notes (2012: 84), first timers are often surprised by how little time in a machine learning project is spent actually doing machine learning. Participants will know that when calculating the highest scoring classifier, more is at stake than just optimization. Coming back to Breiman examples, they will know that at some point doctors or judges will look at these results and eclectically combine them with other sources including their domain expertise to sustain more esoteric claims about diagnosing cancer or reforming the judiciary system. This is the frame that Vollmer calls consumptive, that is, the frame in which people consume, that is, mention, talk about and refer to already-made algorithms or algorithms made by others.

When activity is shifting between reproductive and consumptive utilisation of machine learning, an accordant shift in the framing of activity needs to take place. Goffman calls this transformation a keying (Goffman 1974: 45). Keying informs that a significant shift of frame has occurred, for example, when transforming serious activity into play. This is when even the deepest and most arcane machine learning operations acquire a witness-able nature. Up-keying (i.e. the move from consumptive to reproductive) and down-keying (i.e. the reverse) have to be coordinated across participants. This involves an ability of participants to key both kinds of frames on cue of 'reference-shifting' (Barnes 1983: 534). Goffman else-where uses the notion of 'footing' (Goffman 1981) to highlight for exam-ple how a paediatrician shifts between different footings according to whether he or she is addressing the child, the mother or a video-camera that is recording the consultation for the benefit of a non-present audience of medical students (Tannen and Wallat 1987). Similarly, I envisage a study of deep machine learning whereby the ethnographer investigates how the inner workings of the algorithm are made visible by the pattern-ing of how data scientists talk about them to the screen,[5] to each other and to an audience of funders, domain experts and other interested parties including the broader public.[6]

Algorithms develop in a sequence of familiar set of relations (data sets, organisational work, budgets, priorities and strategic decisions) that enable them to be brought into being and have a predictive effect on human activity. The aim of future ethnographic research on data science should be to expand the algorithmic drama (Ziewitz 2016) in two directions.

First it should encompass the latest developments of machine learning by problematising the 'inner workings' of algorithms. Importantly, due to the dominant realist stance on the notion, current explainability debate is limited to supervised machine learning. Free from the assumption that being accountable implies a mechanistic understanding of the inner func-tioning of the algorithm, it is important that further social research on the interpretability should address deeper forms of machine learning such as reinforcement learning and interactive learning.

Second, future ethnographies should study the continuously evolving accountable order of algorithmic systems along the entire lifecycle of the technological solution *including* the process of its design and testing. This approach should include an attention for the more conceptual level of the so-called deep architectural design of learning as well as the surrounding

---

[5] In a seminal paper on what he called response cries, Goffman (1981: 98) discussed public displays of emotions in the form of vocalisations such as self-talk, imprecations and audible surprise. Goodwin (1996) documented response cries in an operation room at an airport, showing how response cries function as a resource for making co-participants pay attention to something particular on a monitor. A similar attempt in the field of the Sociology of Finance is found in Preda (2009).

[6] A promising way of accessing the sites of algorithmic development and testing is to attend special events such as Hackathons or Data Analytics Fora. I routinely conduct fieldwork at these venues and the level of proximity one can achieve is appropriate to conduct observa-tions such as those outlined in this chapter.

material and social environment as an 'interactional partner for participants' (Knorr Cetina 2009: 72)—see the concluding section of Chap. 2.

Given the direction Machine Learning is taking with respect to the design of the interactions of networks or decision trees in multi-layered machine learning architectures (Domingos 2015), scholars interested in the study of the algorithmic, 'post-social' interactional order (Knorr Cetina and Bruegger 2002) should devote more and more attention to how algorithms interact with each other (MacKenzie 2019) as part of the chain of negotiations that make machine learning accountable.

References

Barnes, B. (1983). Social Life as Bootstrapped Induction. *Sociology, 17*, 524–545.

Bengio, Y., Courville, A., & Vincent, P. (2013). Representation Learning: A Review and New Perspectives. *IEEE Transactions on Pattern Analysis and Machine Intelligence, 35*(8), 1798–1828.

Breiman, L. (2011). Statistical Modeling: The Two Cultures. *Statistical Science, 16*(3), 199–231.

Cardon, D., Cointet, J. P., & Mazières, A. (2018). Neurons Spike Back. The Invention of Inductive Machines and the Artificial Intelligence Controversy. *Réseaux, 211*(5), 173–220.

Collins, H. M. (1990). *Artificial Experts: Social Knowledge and Intelligent Machines.* Cambridge, MA: MIT Press.

Diakopoulos, N. (2016). Accountability in Algorithmic Decision-Making: A View from Computational Journalism. *Acmqueue, 13*(9), 1–24.

Domingos, P. (2012). A Few Useful Things to Know About Machine Learning. *Communications of the ACM., 55*(10), 78–87.

Domingos, P. (2015). *The Master Algorithm: How the Quest for the Ultimate Learning Machine Will Remake Our World.* New York: Basic Books.

Dourish, P., & Button, G. (1998). On "Technomethodology": Foundational Relationships Between Ethnomethodology and System Design. *Human-Computer Interaction, 13*(4), 395–432.

Dreyfus, H. L. (1992). Response to Collins, Artificial Experts. *Social Studies of Science, 22*(4), 717–726.

Epstein, Z., et al. (2018). *Closing the AI Knowledge Gap.* Preprint. Retrieved from https://arxiv.org/abs/1803.07233.

Fitzgerald, D., & Callard, F. (2015). Social Science and Neuroscience Beyond Interdisciplinarity: Experimental Entanglements. *Theory, Culture & Society, 32*(1), 3–32.

Garfinkel, H. (1967). *Studies in Ethnomethodology.* Englewood Cliffs, NJ: Prentice-Hall.

Goffman, E. (1963). *Behavior in Public Places: Notes on the Social Organization of Gatherings.* New York: The Free Press.

Goffman, E. (1974). *Frame Analysis: An Essay on the Organization of Experience.* Boston: Northeastern University Press.

Goffman, E. (1981). *Forms of Talk.* Philadelphia: University of Pennsylvania Press.

Goodwin, C. (1996). Transparent Vision. In E. Ochs, S. Emanuel, & S. Thompson (Eds.), *Interaction and Grammar* (pp. 370–404). Cambridge: Cambridge University Press.

Grudin, J. (2009). AI and HCI: Two Fields Divided by a Common Focus. *AI Magazine, 30*(4), 48–57.

Heath, C., & Button, G. (2002). Editorial Introduction to Special Issue on Workplace Studies. *British Journal of Sociology, 53*(2), 157–161.

Introna, L. (2016). Algorithms, Governance, and Governmentality: On Governing Academic Writing. *Science, Technology & Human Values, 41*(1), 17–49.

Jaton, F. (2017). We Get the Algorithms of Our Ground Truths: Designing Referential Databases in Digital Image Processing. *Social Studies of Science, 47*(6), 811–840.

Knorr Cetina, K. (2009). The Synthetic Situation: Interactionism for a Global World. *Symbolic Interaction, 32*(1), 61–87.

Knorr Cetina, K., & Bruegger, U. (2002). Global Microstructures: The Virtual Societies of Financial Markets. *American Journal of Sociology, 107*, 905–951.

Latour, B., Jensen, P., Venturini, T., Grauwin, S., & Bouillier, D. (2012). 'The Whole Is Always Smaller Than Its Parts'—A Digital Test of Gabriel Tardes' Monads. *The British Journal of Sociology, 63*(4), 590–615.

Lipton, Z. (2018). The Mythos of Model Interpretability. *Communications of the ACM, 61*(10), 36–43.

MacKenzie, A. (2017). *Machine Learners. Archaeology of a Data Practice.* Cambridge MA: The MIT Press.

MacKenzie, D. (2019). How Algorithms Interact: Goffman's 'Interaction Order' in Automated Trading. *Theory, Culture & Society, 36*(2), 39–59.

Markoff, J. (2015). *Machines of Loving Grace. Between Human and Robots.* New York: HarperCollins Publishers.

McGoey, L. (2012). Strategic Unknowns: Towards a Sociology of Ignorance. *Economy and Society, 41*(1), 1–16.

Muniesa, F. (2011). Is a Stock Exchange a Computer Solution? Explicitness, Algorithms and the Arizona Stock Exchange. *International Journal of Actor-Network Theory and Technological Innovation, 3*(1), 1–15.

Neyland, D. (2016). Bearing Account-able Witness to the Ethical Algorithmic System. *Science, Technology, & Human Values, 41*(1), 50–76.

Parrika, J. (2015). *A Geology of Media.* Minneapolis, MN: University of Minnesota Press.

Preda, A. (2009). Brief Encounters: Calculation and the Interaction Order of Anonymous Electronic Markets. *Accounting, Organizations and Society, 34*, 675–693.

Rahwan, I., Cebrian, M., Obradovich, N. M., Bongard, J., Bonnefon, J. F., Breazeal, C., Crandall, J. W., Christakis, N. A., Couzin, I. D., Jackson, M. O., Jennings, N. R., KAmar, E., Kloumann, I. M., Larochelle, H., Lazer, D. R. M. E., Mislove, A., Parkes, D. C., Pentland, A., Roberts, M. E., Shariff, A., Tenenbaum, J. B., & Wellman, M. (2019). Machine Behaviour. *Nature, 568*, 477–486.

Ribeiro, M. T., Singh, S., & Guestrin, C. (2016). *'Why Should I Trust You?' Explaining the Predictions of Any Classifier*. In Proceedings of the 22nd SIGKDD Intern. Conf. Knowledge Discovery and Data Mining, pp. 1135–1144.

Schutz, A. (1971). The Problem of Social Reality. In N. Maurice (Ed.), *Collected Papers I*. the Hague: Nijhoff.

Suchman, L. A. (1987). *Plans and Situated Actions: The Problem of Human-Machine Communication*. Cambridge: Cambridge Press.

Suchman, L., Trigg, R., & Blomberg, J. (2002). Working Artefacts: Ethnomethods of the Prototype. *British Journal of Sociology, 53*(2), 163–179.

Tannen, E., & Wallat, C. (1987). Interactive Frames and Knowledge Schemas in Interaction: Examples from a Medical Examination/Interview. *Social Psychology Quarterly, 50*, 205–216.

Vera, A. H., & Simon, H. A. (1993). Situated Action: A Symbolic Interpretation. *Cognitive Science, 1*, 7–48.

Vollmer, H. (2007). How to Do More with Numbers: Elementary Stakes, Framing, Keying and the Three-Dimensional Character of Numerical Signs. *Accountaing, Organizations and Society, 32*, 577–600.

Voosen, P. (2017). The AI Detectives. *Science, 357*, 22–27.

Watcher, S., Mittelstadt, B., & Russel. C. (2020). *Why Fairness Cannot Be Automated: Bridging the Gap Between EU Non-Discrimination Law and AI*. Preprint. Retrieved from https://arxiv.org/abs/2005.05906.

Ziewitz, M. (2016). Special Issue Introduction: Governing Algorithms: Myth, Mess, and Methods. *Science, Technology, & Human Values, 41*(1), 3–16.

References

Abbott, A. (1988). *The System of Professions. An Essay on the Division of Expert Labor*. Chicago: The University of Chicago Press.

Abbott, A. (1992). From Causes to Events: Notes of Narrative Positivism. *Sociological Methods & Research, 20*(4), 428–455.

Abbott, A. (1995). Sequence Analysis: New Methods for Old Ideas. *Annual Review of Sociology, 21*, 93–113.

Abbott, A. (1997). Of Time and Space: The Contemporary Relevance of the Chicago School. *Social Forces., 75*(4), 1149–1182.

Abbott, A. (2005). Linked Ecologies: States and Universities as Environments for Professions. *Sociological Theory, 23*(3), 245–274.

Anderson, C. (2008). *The End of Theory: The Data Deluge Makes the Scientific Method Obsolete*. Wired Magazine. Retrieved from 23 June, 2008, from https://www.wired.com/2008/06/pb-theory/.

Barad, K. (2007). *Meeting the Universe Halfway: Quantum Physics and the Entanglement of Matter and Meaning*. Durham, NC: Duke University Press.

Barnes, B. (1983). Social Life as Bootstrapped Induction. *Sociology, 17*, 524–545.

Barry, A., & Born, G. (2014). *Interdisciplinarity: Reconfigurations of the Social and Natural Sciences*. London: Routledge.

Barry, A., Born, G., & Weszkalnys, G. (2008). Logics of Interdisciplinarity. *Economy and Society, 37*(1), 20–49.

Bartlett, A., Lewis, J., Reyes-Galindo, L., & Stephens, N. (2018). The Locus of Legitimate Interpretation in Big Data Sciences: Lessons for Computational Social Science from -Omic Biology and High-Energy Physics. *Big Data & Society, 5*(1), 1–15.

© The Author(s), under exclusive license to Springer Nature Switzerland AG 2020

G. Campagnolo, *Social Data Science Xennials*, https://doi.org/10.1007/978-3-030-60358-8

Bateson, G. (2002 [1979]). *Mind and Nature: A Necessary Unity*. Cresskill, NJ: Hampton Press.

Beaulieu, A. (2016). Vectors for Fieldwork: Computational Thinking and New Modes of Ethnography. In L. Hjorth, H. Horst, A. Galloway, & G. Bel (Eds.), *In Companion to Digital Ethnography* (pp. 29–39). London: Routledge.

Becker, H. (1986). *Writing for Social Scientists: How to Start and Finish Your Thesis, Book, or Article*. Chicago: University of Chicago Press.

Bengio, Y., Courville, A., & Vincent, P. (2013). Representation Learning: A Review and New Perspectives. *IEEE Transactions on Pattern Analysis and Machine Intelligence, 35*(8), 1798–1828.

Berger, G., Gan, L., & Fritzler, A. (2016). *How to Become and Executive*. Retrieved from https://www.linkedin.com/pulse/how-become-executive-guy-berger-ph-d-?published=t.

Bergmann, L. (2016). Toward Speculative Data: "Geographic Information" for Situated Knowledges, Vibrant Matter, and Relational Spaces. *Society and Space, 34*(6), 971–989.

Bidwell, M., & Briscoe, F. (2010). The Dynamics of Interorganizational Careers. *Organization Science, 21*(5), 1034–1053.

Bittner, E. (1965). The Concept of Organization. *Social Research, 32*(3), 239–255.

Blok, A., & Pedersen, M. A. (2014). Complementary Social Science? Quali-Quantitative Experiments in a Big Data World. *Big Data & Society, 1*(2), 1–6.

Bouillier, D. (2018). Médialab Stories: How to Align Actor Network Theory and Digital Methods. *Big Data & Society, 5*(2), 1–13.

Bourdieu, P. (1977). *Outline of a Theory of Practice*. Cambridge: Cambridge University Press.

Boyer, D. (2014). Reflexivity Reloaded: From Anthropology to Intellectuals Critique of Method to Studying Sideways. In T. H. Eriksen et al. (Eds.), *Anthropology Now and Next : Essays in Honor of Ulf Hannerz*. New York: Berghahn Books, Incorporated.

Breiman, L. (2011). Statistical Modeling: The Two Cultures. *Statistical Science, 16*(3), 199–231.

Burnap, P., Rana, O., Williams, M., et al. (2014). COSMOS: Towards an Integrated and Scalable Service for Analyzing Social Media on Demand. *International Journal of Parallel, Emergent and Distributed Systems (IJPEDS), 30*(2), 80–100.

Burrows, R., & Gane, N. (2006). Geodemographics, Software and Class. *Sociology, 40*(5), 793.

Callon, M. (1986). Power, Action and Belief: A New Sociology of Knowledge? In J. Law (Ed.), *Some Elements of a Sociology of Translation: Domestication of the Scallops and the Fishermen of St Brieuc Bayin* (pp. 196–223). London: Routledge.

Campagnolo, G. M. (2013). The Evolution of Client-Consultant Relationships: A Situational Analysis of IT Consultancy in the Public Sector. *Financial Accountability & Management, 29*(2), 0267–4424.

Campagnolo, G. M., & Fele, G. (2010). From Specifications to Specific Vagueness: How Enterprise Software Mediates Engineering Relations. *Engineering Studies, 2*(3), 221–243.

Campagnolo, G. M., Pollock, N., & Williams, R. (2015). Technology as We Do Not Know It: The Extended Practice of Global Software Development. *Information and Organization, 25*, 150–159.

Campagnolo, G. M., Nguyen, H., & Williams, R. (2019). The Temporal Dynamics of Technology Promises in Government and Industry Partnerships for Digital Innovation: The Case of the Copyright Hub. *Technology Analysis & Strategic Management, 31*(8), 972–985.

Cardon, D., Cointet, J. P., & Mazières, A. (2018). Neurons Spike Back. The Invention of Inductive Machines and the Artificial Intelligence Controversy. *Réseaux, 211*(5), 173–220.

Castelle, M. (2018). *The Social Lves of Generative Adversarial Networks.* Conference on Fairness, Accountability and Transparency (FAT* '20) January 27–30, 2020, Barcelona, Spain. ACM, New York, NY.

Castells, M. (2000). *The Rise of the Network Society.* Oxford: Blackwell.

Ciborra, C. U. (2000). *From Control to Drift: The Dynamics of Corporate Information Infrastructures.* Oxford: Oxford University Press.

Ciborra, C. (2006). The Mind or the Heart? It Depends on the (Definition of) Situation. *Journal of Information Technology, 21*, 129–139.

Clarke, A. E. (2005). *Situational Analysis: Grounded Theory After the Postmodern Turn.* Thousand Oaks, CA: SAGE.

Clifford, J., & Marcus, G. E. (Eds.). (1986). *Writing Culture: The Poetics and Politics of Ethnography.* Berkeley: University of California Press.

Coleman, G. (2012). *Coding Freedom: The Ethics and Aesthetics of Hacking.* Princeton: Princeton University Press.

Collins, R. (1984). Statistics Versus Words. *Sociological Theory, 2*, 329–362.

Collins, H. M. (1990). *Artificial Experts: Social Knowledge and Intelligent Machines.* Cambridge, MA: MIT Press.

Collins, R. (1994). Why the Social Sciences Won't Become High-Consensus, Rapid-Discovery Science. *Sociological Forum, 9*(2), 155–177.

Collins, H. M., & Evans, R. (2007). *Rethinking Expertise.* Chicago, IL: University of Chicago Press.

Coulter, J. (1996). Human Practices and the Observability of the 'Macrosocial'. *Zeitschrift für Soziologie, 25*, 337–345.

Dalton, C., Taylor, L., & Thatcher, J. (2016). *Critical Data Studies: A Dialog on Data and Space.* SSRN. Retrieved from https://ssrn.com/abstract=2761166.

Daston, L. (2000). The Coming into Being of Scientific Objects. In L. Daston (Ed.), *Biographies of Scientific Objects* (pp. 1–14). Chicago, IL: Chicago University Press.

Di Maggio, P. (2015). Adapting Computational Text Analysis to Social Science (and Vice Versa). *Big Data & Society, 2*(2), 1–5.

Diakopoulos, N. (2016). Accountability in Algorithmic Decision-Making: A View from Computational Journalism. *Acmqueue, 13*(9), 1–24.

Domingos, P. (2012). A Few Useful Things to Know About Machine Learning. *Communications of the ACM., 55*(10), 78–87.

Domingos, P. (2015). *The Master Algorithm: How the Quest for the Ultimate Learning Machine Will Remake Our World.* New York: Basic Books.

Dourish, P. (2001). *Where the Action Is: The Foundations of Embodied Interaction.* Cambridge, MA: MIT Press.

Dourish, P., & Button, G. (1998). On "Technomethodology": Foundational Relationships Between Ethnomethodology and System Design. *Human-Computer Interaction, 13*(4), 395–432.

Downey, G. L. (2009). What Is Engineering Studies For?: Dominant Practices and Scalable Scholarship. *Engineering Studies, 1*(1), 55–76.

Dreyfis, H. L. (1992). Response to Collins, Artificial Experts. *Social Studies of Science, 22*(4), 717–726.

Duggan, M. (2015). *Mobile Messaging and Social Media 2015.* Pew Research Center. Retrieved from http://www.pewinternet.org/files/2015/08/Social-Media-Update-2015-FINAL2.pdf.

Dutta, S., Wei, D., Yueksel, H., Chen, P., Liu, S., & Varshney, K. R. (2020). *Is There a Trade-Off Between Fairness and Accuracy? A Perspective Using Mismatched Hypothesis Testing.* Proceedings of the 37th International Conference on Machine Learning, Vienna, Austria, PMLR 119.

Dutton, H. W. (2013). The Social Shaping of Digital Research. *International Journal of Social Research Methodology, 16*(3), 177–195.

Edwards, A., Housley, W., Williams, M., Sloan, L., & Williams, M. (2013). Digital Social Research, Social Media and the Sociological Imagination: Surrogacy, Augmentation and Re-orientation. *International Journal of Social Research Methodology, 24*, 313–343.

Ehn, P. (1988). *Work-Oriented Design of Computer Artifacts.* PhD diss., Arbetslivscentrum.

Eisenberg, E. M. (1984). Ambiguity as Strategy in Organizational Communication. *Communication Monographs, 51*, 227–242.

Epstein, Z., et al. (2018). *Closing the AI Knowledge Gap.* Preprint. Retrieved from https://arxiv.org/abs/1803.07233.

Evans, J. A., & Aceves, P. (2016). Machine Translation: Mining Text for Social Theory. *Annual Review Sociology, 42*, 21–50.

Fele, G. (2009). Why is Information System Design Interested in Ethnography? Sketches of an Ongoing Story. *Ethnografia e Ricerca Qualitativa, 1,* 1–38.

Fine, T. (1973). *Theories of Probability: An Examination of Foundations.* New York: Academic Press.

Fitzgerald, D., & Callard, F. (2015). Social Science and Neuroscience Beyond Interdisciplinarity: Experimental Entanglements. *Theory, Culture & Society, 32*(1), 3–32.

Gabadinho, A., Ritschard, G., Mueller, N. S., & Studer, M. (2011). Analyzing and Visualizing State Sequences in R with TraMineR. *Journal of Statistical Software, 40*(4), 1–37.

Garfinkel, H. (1967). *Studies in Ethnomethodology.* Englewood Cliffs, NJ: Prentice-Hall.

Gauthier, J. A., Widmer, E. D., Bucher, P., & Notredame, C. (2010). Multichannel Sequence Analysis Applied to Social Science Data. *Sociological Methodology, 40,* 1–38.

George, G., Haas, M., & Pentland, A. S. (2014). Big Data and Management. *Academy of Management Journal, 57*(2), 321–326.

Gillespie, T. (2010). The Politics of Platforms. *New Media and Society, 12*(3), 347–364.

Glaser, B. G., & Strauss, A. L. (1967). *The Discovery of Grounded Theory: Strategies for Qualitative Research.* Chicago: Aldine Pub. Co.

Goffman, E. (1963). *Behavior in Public Places: Notes on the Social Organization of Gatherings.* New York: The Free Press.

Goffman, E. (1974). *Frame Analysis: An Essay on the Organization of Experience.* Boston: Northeastern University Press.

Goffman, E. (1981a). *Forms of Talk.* Philadelphia: University of Pennsylvania Press.

Goffman, E. (1981b). Footing. In E. Goffman (Ed.), *Forms of Talk* (pp. 124–159). Oxford: Blackwell.

Goodwin, C. (1996). Transparent Vision. In E. Ochs, S. Emanuel, & S. Thompson (Eds.), *Interaction and Grammar* (pp. 370–404). Cambridge: Cambridge University Press.

Goulden, M., Greiffenhagen, C., Crowcroft, J., McAuley, D., Mortier, R., Radenkovic, M., & Sathiasaleen, A. (2017). Wild Interdisciplinarity: Ethnography and Computer Science. *International Journal of Social Research Methodology, 20*(2), 137–150.

Gouldner, A. W. (1970). *The Coming Crisis of Western Sociology.* New York: Basic Books.

Greiffenhagen, C., Mair, M., & Sharrock, W. (2011). From Methodology to Methodography: A Study of Qualitative and Quantitative Reasoning in Practice. *Methodological Innovations Online, 6*(3), 93–107.

Grudin, J. (2009). AI and HCI: Two Fields Divided by a Common Focus. *AI Magazine, 30*(4), 48–57.

Guillaume, J., & Latapy, M. (2006). Bipartite Graphs as Models of Complex Networks. *Physica A: Statistical and Theoretical Physics, 371*(2), 795–813.

Hacking, I. (1975). *The Emergence of Probability*. Cambridge: Cambridge University Press.

Hadi, D., & Marcus, G. E. (2011). In the Green Room: An Experiment in Ethnographic Method at the WTO. *PoLAR, 34*(1), 51–76.

Halavais, A. (2013). Structure of Twitter: Social and Technical. In K. Weller, A. Bruns, J. Burgess, M. Mahart, & C. Puschmann (Eds.), *Twitter and Society* (pp. 29–42). New York: Peter Lang.

Halford, S., & Savage, M. (2017). Speaking Sociologically with Big Data: Symphonic Social Science and the Future for Big Data Research. *Sociology, 51*(6), 1–18.

Halfpenny, P., & Procter, R. (2015). *Innovations in Digital Research Methods*. London: Sage.

Hanseth, O., & Bygstad, B. (2012). *ICT Architecture and Project Risk in Inter-Organizational Settings*. ECIS 2012 Proceedings. 130.

Harper, R., & Carter, K. (1994). Keeping People Apart: A Research Note. *Computer Supported Cooperative Work, 2*(3), 199–207.

Healy, K., & Moody, J. (2014). Data Visualization in Sociology. *Annual Review Sociology, 40*, 105–128.

Heath, C., & Button, G. (2002). Editorial Introduction to Special Issue on Workplace Studies. *British Journal of Sociology, 53*(2), 157–161.

Hindess, B. (1973). *The Use of Official Statistics in Sociology: A Critique of Positivism and Ethnomethodology*. London: Macmillan.

Hoffman, A. (2004). Reconsidering the Role of the Practical Theorist: On (re) Connecting Theory to Practice in Organization Theory. *Strategic Organization, 2*(2), 213–222.

Holmes, D., & Marcus, G. (2005). Cultures of Expertise and the Management of Globalisation: Towards the re-functioning of Ethnography. In A. Ong & S. Collier (Eds.), *Global Assemblages: Technology, Politics and Ethics as Anthropological Problems* (pp. 235–252). Malden, MA: Blackwell.

Housley, W., & Smith, R. J. (2017). Interactionism and Digital Society. *Qualitative Research, 17*(2), 187–201.

Housley, W., Procter, R., Edwards, A., Burnap, P., Williams, M., Sloan, L., & Greenhill, A. (2014). Big and Broad Data and the Sociological Imagination: A Collaborative Response. *Big Data and Society, 1*(2), 1–15.

Hutchins, E. (1996). *Cognition in the Wild*. Cambridge: MIT Press.

Hutchins, E., & Palen, L. (1998). Constructing Meaning from Space, Gesture and Speech. In Resnick, L., Säljö, R., Pontecorvo, C. & Burge, B. (eds) 1997. Discourse, Tools and Reasoning: Essays on Situated Cognition. Berlin: Springer-Verlag.

Hyysalo, S. (2010). *Health Technology Development and Use: From Practice-Bound Imagination to Evolving Impacts*. London: Taylor & Francis.

Introna, L. (2016). Algorithms, Governance, and Governmentality: On Governing Academic Writing. *Science, Technology & Human Values, 41*(1), 17–49.

Irons, R. L. (1998). Organizational and Technical Communication: Terminological Ambiguity in Representing Work. *Management Communication Quarterly, 12*(1), 42–71.

Jaton, F. (2017). We Get the Algorithms of Our Ground Truths: Designing Referential Databases in Digital Image Processing. *Social Studies of Science, 47*(6), 811–840.

Kitchin, R. (2014). Big Data, New Epistemologies and Paradigm Shifts. *Big Data & Society, 1*(1), 1–12.

Kitchin, R., & McArdle, G. (2016). What Makes Big Data, Big Data? Exploring the Ontological Characteristics of 26 Datasets. *Big Data & Society, 3*(1). https://doi.org/10.1177/2053951716631130.

Knorr Cetina, K. (1997). Sociality with Objects: Social Relations in Postsocial Knowledge Societies. *Theory, Culture & Society, 14*(4), 1–30.

Knorr Cetina, K. (2009). The Synthetic Situation: Interactionism for a Global World. *Symbolic Interaction, 32*(1), 61–87.

Knorr Cetina, K., & Bruegger, U. (2002). Global Microstructures: The Virtual Societies of Financial Markets. *American Journal of Sociology, 107*, 905–951.

Knorr Cetina, K. (1999). *Epistemic Cultures: How the Sciences Make Knowledge.* Cambridge, MA: Harvard University Press.

Kunda, G. (1991). *Engineering Culture: Control and Commitment in a High-Tech Corporation.* Philadelphia: Temple University Press.

Latour, B. (1999). *Pandora's Hope: Essays on the Reality of Science Studies.* Cambridge, MA: Harward University Press.

Latour, B. (2005). *Reassembling the Social: An Introduction to Actor-Network-Theory.* Oxford: Oxford University Press.

Latour, B., Jensen, P., Venturini, T., Grauwin, S., & Bouillier, D. (2012). 'The Whole Is Always Smaller Than Its Parts'—A Digital Test of Gabriel Tardes' Monads. *The British Journal of Sociology, 63*(4), 590–615.

Lave, J. (1988). *Cognition in Practice: Mind, Mathematics and Culture in Everyday Life.* New York: Cambridge University Press.

Law, J., & Urry, J. (2004). Enacting the Social. *Economy and Society, 33*(3), 390–410.

Lazer, D., & Radford, J. (2017). Data Ex Machina: Introduction to Big Data. *Annual Review of Sociology, 43*, 19–39.

Leonardi, P. M. (2011). When Flexible Routines Meet Flexible Technologies: Affordance, Constraint, and the Imbrication of Human and Material Agencies. *MIS Quarterly, 35*(1), 147–167.

Lévi-Strauss, C. (1963). *Structural Anthropology.* New York: Doubleday Anchor Books.

Lipton, Z. (2018). The Mythos of Model Interpretability. *Communications of the ACM, 61*(10), 36–43.

Lupton, D. (2014). *Digital Sociology*. London and New York: Routledge.

MacKenzie, D. (1981). *Statistics in Britain, 1865–1930: The Social Construction of Scientific Knowledge*. Edinburgh: Edinburgh University Press.

MacKenzie, A. (2017). *Machine Learners. Archaeology of a Data Practice*. Cambridge MA: The MIT Press.

MacKenzie, D. (2018). 'Making', 'Taking' and the Material Political Economy of Algorithmic Trading. *Economy and Society, 47*(4), 501–523.

MacKenzie, D. (2019). How Algorithms Interact: Goffman's 'Interaction Order' in Automated Trading. *Theory, Culture & Society, 36*(2), 39–59.

Mahoney, J., & Goertz, G. (2006). A Tale of Two Cultures: Contrasting Quantitative and Qualitative Research. *Political Analysis, 14*(3), 33–53.

Marcus, G. (2002). Intimate Strangers: The Dynamics of (Non) Relationship Between the Natural and Human Sciences in the Contemporary U.S. University. *Anthropological Quarterly, 75*(4), 519–526.

Markoff, J. (2015). *Machines of Loving Grace. Between Human and Robots*. New York: HarperCollins Publishers.

Marres, N. (2002). May the True Victim of Defacement Stand Up! On Reading the Network Configurations of Scandal on the Web. In B. Latour & P. Weibel (Eds.), *Iconoclash, Image-Making in Science, Religion and Art*. Karlsruhe/Cambridge: ZKM/MIT Press.

Marres, N. (2017). *Digital Sociology*. Cambridge: Polity Press.

Marres, N., & Moats, D. (2015). *Mapping Controversies with Social Media: The Case for Symmetry*. SSRN. Retrieved from https://ssrn.com/abstract=2567929 or https://doi.org/10.2139/ssrn.2567929.

Marres, N., & Weltevrede, E. (2015). Scraping the Social? Issues in Real-Time Social Research. In J. Bouchard, É. Candel, H. Cardy, & G. Gomez-Mejia (Eds.), *La Médiatisation de l'Évaluation*. Berlin: Peter Lang.

Marrow, A. (1977). *The Practical Theorist: The Life and Work of Kurt Lewin*. New York: Teachers College Press.

McFarland, A. D., Lewis, K., & Goldberg, A. (2015). Sociology in the Era of Big Data: The Ascent of Forensic Social Science. *American Society, 47*, 12–35.

McGoey, L. (2012). Strategic Unknowns: Towards a Sociology of Ignorance. *Economy and Society, 41*(1), 1–16.

Metzler, K. (2016). *"The Big Data Rich and the Big Data Poor": The New Digital Divide Raises Questions About Future Academic Research*. The Impact Blog, London School of Economics and Political Science. Retrieved from http://blogs.lse.ac.uk/impactofsocialsciences/2016/11/22/the-big-data-rich-and-the-big-data-poor-the-new-digital-divide-raises-questions-about-future-academic-research/.

Mills, C. W. (1959). *The Sociological Imagination*. Oxford: Oxford University Press.

Moats, D., & Borra, E. (2018). Quali-Quantitative Methods Beyond Networks: Studying Information Diffusion on Twitter with the Modulation Sequencer. *Data & Society, 5*(1), 1–17.

Mol, A. (2002). *The Body Multiple: Ontology in Medical Practice*. Durham, NC: Duke University Press.

Molina, M., & Garip, F. (2019). Machine Learning for Sociology. *Annual Review Sociology, 45*, 27–45.

Monteiro, E., Jarulaitis, G., & Hepsø, V. (2012). The Family Resemblance of Technologically Mediated Work Practices. *Information and Organization, 22*(3), 169–187.

Monteiro, E., Pollock, N., Hanseth, O., & Williams, R. (2013). From Artefacts to Infrastructure. *Computer Supported Cooperative Work, 22*(4–6), 575–607. (CSCW).

Muniesa, F. (2011). Is a Stock Exchange a Computer Solution? Explicitness, Algorithms and the Arizona Stock Exchange. *International Journal of Actor-Network Theory and Technological Innovation, 3*(1), 1–15.

Munk, A., & Jensen, T. E. (2014). Revisiting the Histories of Mapping: Is there a Future for a Cartographic Ethnology? *Ethnologia Europea, 44*(2), 31–47.

Neyland, D. (2016). Bearing Account-able Witness to the Ethical Algorithmic System. *Science, Technology, & Human Values, 41*(1), 50–76.

Nicolini, D. (2011). Practice as the Site of Knowing. Insights from the Field of Telemedicine. *Organization Science, 22*, 602–620.

O'Reilly, K. (2009). *Key Concepts in Ethnography*. London: Sage.

Orr, J. (1996). *Talking About Machines: An Ethnography of a Modern Job*. Ithaca: Cornell University Press.

Orton-Johnson, K., & Prior, N. (Eds.). (2013). *Digital Sociology: Critical Perspectives*. London: Palgrave Macmillan.

Parrika, J. (2015). *A Geology of Media*. Minneapolis, MN: University of Minnesota Press.

Pedersen, M. A., Albris, K., Munch Gregersen, E., & Otto, E. I. (2020). *Machine Anthropology. A Two Day Interdisciplinary Workshop*. Copenhagen: Centre for Social Data Science (SODAS).

Pentland, B. T., & Feldman, M. S. (2008). Designing Routines: On the Folly of Designing Artifacts, While Hoping for Patterns of Action. *Information and Organization, 18*(4), 235–250.

Pollock, N., & Williams, R. (2008). *Software and Organisations: The Biography of the Enterprise-Wide System or How SAP Conquered the World*. Oxon: Routledge.

Pollock, N., & Williams, R. (2016). *How Industry Analysts Shape the Digital Future*. Oxford: Oxford University Press.

Pollock, N., Williams, R., & D'Adderio, L. (2007). Global Software and Its Provenance: Generification Work in the Production of Organizational Software Packages. *Social Studies of Science, 37*, 254–280.

Preda, A. (2009). Brief Encounters: Calculation and the Interaction Order of Anonymous Electronic Markets. *Accounting, Organizations and Society, 34,* 675–693.

Procter, R., Vis, F., & Voss, A. (2013). Reading the Riots on Twitter: Methodological Innovation for the Analysis of Big Data. *International Journal of Social Research Methodology, 16*(3), 197–214.

Radin, P. (1927). *Primitive Man as Philosopher.* New York: D. Appleton and Company.

Rahwan, I., Cebrian, M., Obradovich, N. M., Bongard, J., Bonnefon, J. F., Breazeal, C., Crandall, J. W., Christakis, N. A., Couzin, I. D., Jackson, M. O., Jennings, N. R., KAmar, E., Kloumann, I. M., Larochelle, H., Lazer, D. R. M. E., Mislove, A., Parkes, D. C., Pentland, A., Roberts, M. E., Shariff, A., Tenenbaum, J. B., & Wellman, M. (2019). Machine Behaviour. *Nature, 568,* 477–486.

Randall, D., Harper, R., & Rouncefield, M. (2005). Fieldwork, Ethnography and Design: A Perspective from CSCW. In K. Anderson & T. Lovejoy (Eds.), *EPIC 2005: Ethnographic Praxis in Industry Conference* (pp. 88–99). Seattle, WA and Arlington, VA: Redmond, American Anthropological Association.

Rheinberger, H.-J. (2011). Consistency from the Perspective of an Experimental Systems Approach to the Sciences and Their Epistemic Objects. *Manuscrito, 34*(1), 307–321.

Ribeiro, M. T., Singh, S., & Guestrin, C. (2016). *'Why Should I Trust You?' Explaining the Predictions of Any Classifier.* In Proceedings of the 22nd SIGKDD Intern. Conf. Knowledge Discovery and Data Mining, pp. 1135–1144.

Rieder, B. (2013). *Studying Facebook via Data Extraction: The Netvizz Application.* Proceedings of WebSci '13, the 5th Annual ACM Web Science Conference, pp. 346–355.

Rogers, R. (2013). Internet Research: The Question of Method. In C. Hine (Ed.), *Virtual Research Methods* (pp. 213–238). London: Sage.

Rogers, R., & Marres, N. (2000). Landscaping Climate Change: A Mapping Technique for Understanding Science and Technology Debates on the World Wide Web. *Public Understanding of Science, 9*(2), 141–163.

Salganik, M. J. (2018). *Bit by Bit: Social Research in the Digital Age.* Princeton, NJ: Princeton University Press.

Sandvig, C., & Hargittai, E. (2015). How to Think about Digital Research. In E. Hargittai & C. Sandvig (Eds.), *Digital Research Confidential: The Secrets of Studying Behavior Online.* Cambridge, MA: MIT Press.

Savage, M. (2009). Against Epochalism: An Analysis of Conceptions of Change in British Sociology. *Cultural Sociology, 3*(2), 217–238.

Savage, M. (2015). Sociology and the Digital Challenge. In P. Halfpenny & R. Procter (Eds.), *Innovations in Digital Research Methods.* London: Sage.

Savage, M., & Burrows, R. (2007). The Coming Crisis of Empirical Sociology. *Sociology, 41*(5), 885–899.

Sayer, A. (1992). *Method in Social Science: A Realist Approach* (2nd ed.). London: Routledge.

Schutz, A. (1971). The Problem of Social Reality. In N. Maurice (Ed.), *Collected Papers I*. the Hague: Nijhoff.

Seaver, N. (2014). *Computers and Sociocultural Anthropology*. Savage Minds. Retrieved from https://savageminds.org/2014/05/19/computers-and-sociocultural-anthropology.

Seaver, N. (2017). Algorithms as Culture: Some Tactics for the Ethnography of Algorithmic Systems. *Big Data & Society, 4*(2), 1–12.

Shaw, R. (2015). Big Data and Reality. *Big Data and Society, 2*(2), 1–4.

Snow, C. P. (1959). *The Two Cultures and the Scientific Revolution*. Cambridge: Cambridge University Press.

Sommerville, I, Rodden, T., Sawyer, P., Bentley, R., & Twidale, M. (1993). *Integrating Ethnography Into the Requirements Engineering Process*. In Proc. IEEE International Symposium on Requirements Engineering (RE'93), San Diego, pp. 165–173.

Stengers, I. (2011, December 13). *"Another Science is Possible" A Plea for Slow Science*. Faculté de Philisophie et Lettres, ULB. Inaugural Lecture Chair Williy Calewaert 2011–2012 (VUB).

Suchman, L. A. (1987). *Plans and Situated Actions: The Problem of Human-Machine Communication*. Cambridge: Cambridge Press.

Suchman, L. (1994). Working Relations of Technology Production and Use. *Computer Supported Cooperative Work, 2*(1-2), 21–39.

Suchman, L., Trigg, R., & Blomberg, J. (2002). Working Artefacts: Ethnomethods of the Prototype. *British Journal of Sociology, 53*(2), 163–179.

Tannen, E., & Wallat, C. (1987). Interactive Frames and Knowledge Schemas in Interaction: Examples from a Medical Examination/Interview. *Social Psychology Quarterly, 50*, 205–216.

Tarde, G. (1895). *La logique sociale*. Paris: Félix Alcan.

Turner, S. (2001). What is the Problem with Experts? *Social Studies of Science, 31*(1), 123–149.

Vaast, E., & Walsham, G. (2009). Trans-Situated Learning: Supporting a Network of Practice with an Information Infrastructure. *Information Systems Research, 20*(4), 547–564.

Van Dijck, J. (2012). *The Network Society*. London: Sage.

Veltri, A. G. (2019). *Digital Social Research*. Cambridge: Polity Books.

Venturini, T., Jacomy, M., & Meaner, A. (2017). An Unexpected Journey: A Few Lessons from Sciences Po Médialab's Experience. *Big Data & Society, 4*(2), 1–11.

Venturini, T., Munk, A., & Jacomy, M. (2019). Actor-Network VS Network Analysis VS Digital Networks Are We Talking About the Same Networks? In D. Ribes & J. Vertesi (Eds.), *DigitalSTS: A Handbook and Fieldguide* (pp. 510–523). Princeton, NJ: Princeton University Press.

Vera, A. H., & Simon, H. A. (1993). Situated Action: A Symbolic Interpretation. *Cognitive Science, 1*, 7–48.

Vollmer, H. (2007). How to Do More with Numbers: Elementary Stakes, Framing, Keying and the Three-Dimensional Character of Numerical Signs. *Accountaing, Organizations and Society, 32*, 577–600.

Voosen, P. (2017). The AI Detectives. *Science, 357*, 22–27.

Watcher, S., Mittelstadt, B., & Russel. C. (2020). *Why Fairness Cannot Be Automated: Bridging the Gap Between EU Non-Discrimination Law and AI*. Preprint. Retrieved from https://arxiv.org/abs/2005.05906.

Williams, R., & Edge, D. (1996). The Social Shaping of Technology. *Research Policy Vol., 25*(1996), 856–899.

Williams, R., & Procter, R. (1998). Trading Places: A Case Study of the Formation and Deployment of Computing Expertise. In R. Williams et al. (Eds.), *Exploring Expertise* (pp. 197–222). Basingstoke: Macmillan. Chap. 13.

Wilson, E. (1998). *Consilience:The Unity of Knowledge*. New York: Knopf.

Index[1]

A

Abduction, 53, 61n4
Accountability, 13, 30, 74,
76, 80–85
Agonistic-antagonistic
mode, 7, 52
Alan Turing Institute (ATI), 8
Algorithm
algorithmic drama, 87
algorithmic naturalness, 79
Ambiguity, 10, 20, 23, 36, 42
Appresentation, 24–30, 84
Artificial Intelligence (AI),
13, 24, 30n6, 75–78,
80–83, 84n3

B

Benign view, 3
Black box, 12, 74, 79

C

Careers, 4, 6, 9, 13, 46, 55–61,
63–69, 68n12
Classifier, 74, 81, 84, 86
Commercial sociology, 29, 62
Computer science, 3, 5–8, 20,
23, 52, 80
Confidential, 4, 51
Connectionism, 13, 75–77, 76n1
Consilience, 53, 54, 61n4
Contributory expertise, 11
Controversy Mapping, 5

D

Data pre-processing, 56, 66
Decision trees, 84, 84n3, 85, 88
Deep learning, 13, 47, 75, 77, 78,
82, 85, 87
Deflationary sensibility, 13

---

[1] Note: Page numbers followed by 'n' refer to notes.

Digital bandwagon, 5
Digitally produced network graphs, 5
Digital social research (DSR), 2–6,
    8n9, 9, 12, 24, 36, 43, 53
    lack of ambition, 12, 36
Digital STS, 6n6
Disciplinary divide, 8
Disciplinary politics, 52
Disunity, 3–4

E
Edinburgh-centric perspective, 5
Editorial politic, 4
Empractised, 48, 53
Epistemic cultures, 13, 58
Epochalist thinking, 48
Ethnography, 6–8, 10, 13, 20, 21, 32,
    53, 54, 56–58, 61, 62, 68,
    69, 73–88
    ethnographic approach, 36,
        62, 67, 68
Ethnomethodology, 4, 8–13, 20,
    20n1, 22, 23n2, 24, 25, 29–31,
    30n5, 30n6, 77, 81
Experimental entanglements, 62
Explainable AI (xAI), 13, 77, 83–86

F
Fairness, 81
Feature engineering, 74
First-person perspective, 4
Footing, 87
Formative effect argument, 29

G
Gatekeepers, 68
Genealogy, 3, 5, 48
Golden standard, 64
Ground truth, 13, 74

H
Human-computer interaction (HCI),
    24, 80, 82

I
Ignorance, 84
Informants, 55, 56, 63,
    67, 68
Infrastructure studies, 8–10, 24
Integration of social facts, 31
Intellectual oblivion, 3
Interactionism, 4, 11, 20, 20n1,
    24, 28, 77
Interpretable AI, 77, 83
IssueCrawler, 5

K
Keying, 86

L
Linked ecology approach, 9, 12,
    38, 41–43
Localist turn, 30
Locus of legitimate
    interpretation, 7, 66
Logistic regression, 85
Low fidelity, 45

M
Machine learning, 8, 10,
    11, 13, 47, 64, 73–75, 77,
    79–88, 85n4
Mathematical problem solving,
    10, 20, 21
Media materialism, 74
Mixed methods, 53, 54
Model ensembles, 13
Modifications, 7, 8, 53

N

Narrative positivism, 3, 52
Network approaches, 5, 10, 36, 45
Neural networks, 76, 77, 79
New AI, 76
New research hardware, 48
Non-positive quantitative approaches, 36
Non-positivist frame, 52
Normalisation, 65

O

Old theories, 36, 46–48, 52
Ontologically fluid, 31
Opacity, 12
Optimisation, 77, 78, 86

P

Para-ethnography, 62–64
Para-site, 9, 63, 64
Participative epistemology, 51–69
Personal intellectual biography, 10
Phenomenology, 7
    phenomenological sociology, 3, 7–8, 10, 12, 13, 20, 25, 35, 52, 74, 75, 77, 78, 81, 85
Practical theorist, 63
Processual
    process analysis, 2n1
    processual approach, 66n9
    processual perspective, 4–6, 64
    processual way, 2

Q

Qualitative social science, 6

R

Random forest, 77, 85
Regime of the inter-, 80

S

Sequence analysis, 6, 9, 11, 13, 52, 54, 57, 58, 61, 62, 66–68
Singular methods, 10
Situated studies, 12, 20, 24
Situational analysis, 9, 12, 38–41, 43
Slow science, 36
Sociological imagination, 11, 46
Socio-materiality, 2n1, 13, 74–75
Statistics
    social statistics, 7–8, 66
    statistically significant, 66, 67
    statisticians, 6, 77
Supervised model, 65
Symbolic AI, 13, 75, 76, 80
Symbolic interactionism, 4, 20n1, 24
Symphonic social science, 53, 54, 61n4

T

Technomethodology, 13, 23n2, 81–83
Text mining, 9, 11, 13, 52, 54, 55, 58, 61
Transparency, 45, 83, 84, 84n3
Turing Box, 78
Twitter, 44
Two cultures, 8, 85

U
Unconscious biases, 29, 62, 66, 80

V
Vagueness, 10, 20–23, 36, 84
Value-neutral, 42

W
Wild interdisciplinarity, 64

X
xAI, *see* Explainable AI
Xennials, 1–13